GENETIC DIAGNOSES

GENETICS - RESEARCH AND ISSUES

Additional books in this series can be found on Nova's website
under the Series tab.

Additional E-books in this series can be found on Nova's website
under the E-books tab.

GENETIC DIAGNOSES

RADHA JONNALAGEDDA SARMA
EDITOR

Nova Science Publishers, Inc.
New York

For permission to use material from this book please contact us:
Telephone 631-231-7269; Fax 631-231-8175
Web Site: http://www.novapublishers.com

NOTICE TO THE READER
The Publisher has taken reasonable care in the preparation of this book, but makes no expressed or implied warranty of any kind and assumes no responsibility for any errors or omissions. No liability is assumed for incidental or consequential damages in connection with or arising out of information contained in this book. The Publisher shall not be liable for any special, consequential, or exemplary damages resulting, in whole or in part, from the readers' use of, or reliance upon, this material. Any parts of this book based on government reports are so indicated and copyright is claimed for those parts to the extent applicable to compilations of such works.

Independent verification should be sought for any data, advice or recommendations contained in this book. In addition, no responsibility is assumed by the publisher for any injury and/or damage to persons or property arising from any methods, products, instructions, ideas or otherwise contained in this publication.

This publication is designed to provide accurate and authoritative information with regard to the subject matter covered herein. It is sold with the clear understanding that the Publisher is not engaged in rendering legal or any other professional services. If legal or any other expert assistance is required, the services of a competent person should be sought. FROM A DECLARATION OF PARTICIPANTS JOINTLY ADOPTED BY A COMMITTEE OF THE AMERICAN BAR ASSOCIATION AND A COMMITTEE OF PUBLISHERS.

Additional color graphics may be available in the e-book version of this book.

Library of Congress Cataloging-in-Publication Data

Genetic diagnoses / editor, Radha Jonnalagedda Sarma.
     p. ; cm.
  Includes bibliographical references and index.
  ISBN 978-1-61324-866-9 (hardcover)
  1. Human chromosome abnormalities--Diagnosis. 2. Molecular biology. I. Sarma, Radha Jonnalagedda.
  [DNLM: 1. Genetic Testing. 2. Genetic Counseling. 3. Genetic Techniques. 4. Molecular Biology. QZ 52]
  RB155.6.G4613 2011
  616'.042--dc23
                          2011018627

Published by Nova Science Publishers, Inc. ✛ New York

Contents

Preface

This book presents current research from across the globe in the study of genetic diagnoses. Topics discussed include the genetic risk factors between thyroid function and preeclampsia; the molecular diagnosis of thalessemia and spinocerebelar ataxias; the genetics of left ventricular noncompaction; the genetic diagnosis to type 1 diabetes and the genetics of congenital heart disease.

Chapter I - Background: The prime for this chapter was a recent study which oppined upon the correlation of preeclampsia and thyroid function and the critisism which followed, both bringing forward, on one hand, the significance of thyroid pertubations in the pathophysiology in question and, on the other hand, the incoherence still prepondarating. The rising contradictions from the addressing studies are of such extent that one cannot help but wonder whether they pertain to the same field at all. Preeclampsia has been termed "the disease of theories" for the numerous theories concerning its aetiology. Willing not to add another study of theoretical contemplations to the pile of obsurity, effort has been made in this work to stay with the facts, with the aim to somewhat elucidate "what leads to what". Studying classic and recent review articles, reference lists of published letters and asking experts in the field, a scheme and routes are concluded and presented with arguements and selected representing literature. Conclusions: Both hypo- and hyper-thyroidism constitute risk factor to preeclampsia. Additionaly, transient hypothyroidism is induced in mother and in utero, for teleological reasons, subject to regulation by an intact hypothalamus-pituitary-thyroid axis. The clinician should expect raised, normal or slightly pathological TSH up to a time relatively close to delivery. Counteracting influences lead to incalculable alterations in T_3, T_4, FT_3 and FT_4 and though a most common pattern coincides with lower FT_3 and raised FT_4, more diversity is allowed.

Chapter II - The thalassemias are a group of inherited hematologic disorders characterized by hemolytic anemia. They are among the most common single gene disorders in the world and are especially frequent in South-East Asia, the Middle East, India, the Mediterranean and Africa. Due to demographic changes in the last few decades, the incidence of thalassemia in North European countries and North America has increased. In normal adults, the major hemoglobin (Hb) consists of two α- and two β-globin chains which are covalently linked to heme, the oxygen-binding group. The two major forms of thalassemia, α- and β-thalassemia, are due to absent or reduced production of α-globin and β-globin chains, respectively. While the majority of α-thalassemia is due to single or double gene deletions in the α-globin gene, β-thalassemia is mostly caused by point mutations in the β-globin gene.

Affected individuals with different mutations suffer from variable degrees of anemia, which can be detected by hematological/biochemical investigations such as complete blood count (CBC), Hb analysis by isoelectric focusing (IEF) or high-performance liquid chromatography (HPLC). Molecular analysis is usually required to confirm the hematological observations and make a definitive diagnosis of the causative mutations to inform management and counseling. Many molecular diagnostic and screening assays have been developed to identify the mutations in the α- and β-globin genes. This article aims to review the various technical platforms, including PCR based methods and high-throughput approaches, and common strategies available for the molecular diagnosis of thalassemia. Also discussed are thalassemia prevention strategies such as prenatal and pre-implantation genetic diagnosis.

Chapter III - This chapter deals with restoration and identification of the causes (diagnoses) through the observed effects (symptoms) on the basis of fuzzy relational calculus and the least squares method. Fuzzy relational calculus plays the central role as a uniform platform for inverse problem resolution on various fuzzy approximation operators. The diagnosis problem, which is based on a cause and effect analysis and abductive reasoning, can be formally described by fuzzy relations and fuzzy IF-THEN rules. In this chapter, the authors propose an approach for building fuzzy systems of diagnosis, which enables solving fuzzy relational equations together with design and tuning of multiple variable linguistic models on the basis of expert and experimental information. The authors suggest some procedures of numerical solution of the fuzzy relational equations using genetic algorithms. The procedures envisage the optimal solution growing from a set of primary variants by extending the simplified solution set to the case of multidimensional fuzzy relational equations. The essence of tuning consists of the selection of such membership functions of the fuzzy terms for the input and output variables (causes and effects) and such «causes-effects» fuzzy relations, which provide minimal difference between theoretical and experimental results of diagnosis. The problem of tuning is formulated as an optimization problem, which is solved by using a genetic algorithm. The efficiency of the proposed models and algorithms is illustrated by the computer experiments and the real examples from technical diagnosis.

Chapter IV - Spinocerebelar ataxias (SCA) are a clinical and genetic heterogeneous group of neurodegenerative disease, affecting 1-4 cases per 100,000 individuals and presenting a wide geographic variation. SCAs are characterized by progressive degeneration of cerebellum and its afferent and efferent connections. The main clinical symptom is the presence of gait and limb ataxia, and most of them are inherited as a dominant trait. Based in clinical manifestations SCAs can be classified in four groups, but based on the genetic molecular defects, there are around 30 different SCAs, nevertheless some patients remain without a molecular diagnosis. Although the first approach is clinic, molecular diagnosis becomes essential, providing an accurate diagnostic and making possible genetic counselling, presymptomatic testing and prenatal diagnosis. It is important to highlight the fact that while a positive test establish a diagnostic, a negative one does not discard it. The main molecular defect found in SCAs is a dynamic mutation. In most of the cases, it corresponds to a CAG expansion in the coding region of the gene, giving a polyQ protein and sharing the particularities of this kind of mutations: late onset disease with variable expression, complete penetrance, anticipation and being the paternal transmission more severe than the maternal one. SCA alleles are classified in four categories normal, intermediate, reduced penetrance and mutated. Individuals carrying the mutation synthesize an aberrant protein, with an acquired gain of function that damages particular group of cells and that leads to cell death. There are other cases of ataxia, such as FXTAS, in which the molecular defect is also a dynamic mutation but localized in the non-coding region of the gene. Finally there are also ataxias caused by rare conventional mutations (missense, deletion, insertion or duplication). An example is SCA11, which is caused by mutations in *TTBK2* gene. In this chapter the authors present our experience in molecular diagnosis of spinocerebellar ataxias. The authors have performed molecular diagnoses of SCA in more than 500 unrelated patients, including confirmation of clinical suspicion, prenatal and presymptomatic diagnosis for SCA1, SCA2, SCA3, SCA6, SCA7, SCA8, DRPLA and FXTAS. On the basis of our results, the authors can conclude that the most common ataxia in the Spanish population is SCA3 (Machado Joseph) followed by SCA2. Finally and due to the fact that the authors are a reference centre for fragile X syndrome, the authors have a wide experience on FXTAS.

Chapter V – Left ventricular noncompaction (LVNC) is still an unclassified cardiomyopathy according to the World Health Organization classification of cardiomyopathies, but in 2006 the American Heart Association classified this entity as a primary cardiomyopathy of genetic origin. LVNC is characterized by a distinctive spongy appearance due to excessive trabeculation of the myocardium and deep intertrabecular recesses that communicate with the left ventricular cavity. Elucidating a genetic basis for the phenotypic expression of LVNC has been difficult, since both familial and sporadic forms have been described. It appears that the pediatric presentation may have a distinct genetic basis as compared to the adult presentation. After the first case of isolated LVNC was reported 25years ago, much has been published about this entity. Even now, there is no consensus on the diagnostic criteria, and the diagnosis is based on the morphological features identified by cardiac imaging studies or at autopsy. Main complications seen in this condition are heart failure thromboembolism and arrhythmias. Association of neurological conditions has been reported with left ventricular hypertrabeculation. Different genes found to be associated with LVNC are (1) taffazin (TAZ), (2) alfa-dystrobrevin (DTNA), (3) Cypher/ZASP (LDB3) (4) lamin A/C (LMNA), (5) SCN5A, (6) MYH7 and (8) MYBPC3. There is also significant overlap in the phenotypes of the genetically mediated cardiomyopathies. Thus, LVNC can occur as LVNC with dilated or hypertrophic cardiomyopathy. Although further research is needed to elucidate the genetic basis of LVNC, at this time it seems clear that there is considerable genetic heterogeneity involved. In three of the largest series of patients, the rate of familial involvement was 18%, 25%, and 33%. In the familial cases, autosomal dominance is more common than sex-linked inheritance. While genetic testing is not routinely recommended at this time, the Heart Failure Society of America practice guidelines recommend clinical screening of all first degree relatives of affected patients for LVNC. Genetic testing is to be considered for affected patients with a firm diagnosis of LVNC to help with family screening. Prognosis seems to be better in the recent publications likely due to improved therapeutic strategies. Standardization of diagnostic criteria is essential to continue further research and reporting results in LVNC. Genetic research still needs to be continued to further our knowledge of this genetic disorder.

Chapter VI - Type 1 Diabetes (T1D) is a complex trait caused by autoimmune destruction of islet beta cells in the pancreas resulting of the interaction between genetic and environmental factors. Despite enormous advances in the study of T1D there is still no cure and both genetic and environmental factors remain undefined. To date the therapy was mainly aimed at controlling hyperglycemia; however, recent therapeutic approaches include both the immunosuppression and immunetolerance for prevention and treatment of T1D. Autoreactive T cells have been proposed as target of this strategy because they are responsible for the autoimmune beta cell destruction. It is well known that preventing or avoiding the secondary effects of these therapies require a precise diagnosis of the autoimmune diseases. Hence, a genetic diagnosis would be suitable in order to prevent T1D and to achieve a correct immunotherapy. The major T1D susceptibility genes, the HLA class II loci (HLA-DRB1 and HLA-DQB1) on chromosome 6p21 act in combination with other non-HLA genes across the genome. Recently, genome-wide association studies have identified over 40 chromosome regions outside the HLA as being associated with T1D. Some of these non-HLA loci have been implicated in other autoimmune diseases like celiac disease, rheumatoid arthritis or multiple sclerosis. The fact that distinct genetic variants are shared by different autoimmune diseases suggests that common immunological mechanisms are involved in the etiology of these autoimmune diseases. The aim of this chapter is to analyze the T1D associated genes that have been replicated in different populations and whether they could lead to the genetic diagnosis. For this purpose the authors performed an analysis of three main established susceptibility genes in a Spanish cohort. The authors found that the cumulative presence of predisposition variants of the mentioned genes increases significantly the risk to disease.

Chapter VII - Congenital heart disease is the most frequent form of major birth defects in newborns affecting close to 1% of newborn babies (8 per 1,000). The etiology is multifactorial, including a genetic basis (causative genes, or interactions between genes and environment) and the influence of non-inherited risk factors (such as multivitamins, maternal illness, drug administration, environmental agents exposure or also maternal and paternal sociodemographic factors). Heart is considered the first functional organ of the embryo. It develops from the mesodermal sheets through the formation of an early linear heart tube which begins to contract at the eight- to nine-somite stage before the formation of the chambers and the conduction system. Recently, several CHDs have been found to be caused by mutations of the genes involved in the heart embryogenesis (*TFAP2B, Tbx1, NKX2-5, NKX2-6, ZFPM2/FOG2, GATA4*). These genes are often investigated as candidate genes if their biological role could be associated to the cardiac defect. Genome wide linkage analysis, candidate gene association studies, RNA expression profiling and resequencing are commonly used techniques to identify genes responsible of a certain cardiovascular disorder. The method of analysis can be selected on the basis of the known information about the disease. As a consequence of the increasing number of congenital/genetic cardiovascular diseases discovered in recent years, the genetic counselling has become very useful to inform the family of the affected subject about the hereditary risk, and to suggest genetic testing for family members. In addition, due to the significant improvement of non invasive imaging techniques (ultrasound imaging) and molecular analysis, prenatal diagnosis is becoming available for many congenital disorders. Chromosomal karyotype with increased band number to identify large chromosome rearrangements, gene or region specific FISH analysis for detecting deletions/insertions or aneuploidies, indirect PCR assays to assess small gene specific mutations or deletions/insertions, or direct sequence analysis for the detection of specific point mutations represent some of the available techniques to detect the disease in an early phase. Examples of genetic proved congenital disorders for which a genetic test is available are DiGeorge syndrome (deletion 22q11, *Tbx1* gene), Williams-Beuren syndrome (microdeletion 7q11.23, gene contiguous syndrome), Alagille syndrome (deletion 20p12,

JAG1 gene), Noonan and LEOPARD syndrome (*PTPN11*, *SOS1, KRAS*, and *RAF-1*), Holt-Oram syndrome (mutations in *TBX5* gene).

In: Genetic Diagnoses
Editor: R. J. Sarma, pp. 1-27

ISBN: 978-1-61324-866-9
© 2012 Nova Science Publishers, Inc.

Chapter I

Defining a Cause:
Result Priority between Preeclampsia and Thyroid Function

Ioannis Katsakoulas *and **Clio P. Mavragani***
School of Medicine, Department of Exper. Physiology
Univ. of Athens, School of Medicine, Athens, Greece.

Abstract

Background: The prime for this chapter was a recent study ([1]) which oppined upon the correlation of preeclampsia and thyroid function and the critisism which followed, both bringing forward, on one hand, the significance of thyroid pertubations in the pathophysiology in question and, on the other hand, the incoherence still prepondarating. The rising contradictions from the addressing studies are of such extent that one cannot help but wonder whether they pertain to the same field at all. Preeclampsia has been termed "the disease of theories" for the numerous theories concerning its aetiology. Willing not to add another study of theoretical contemplations to the pile of obsurity, effort has been made in this work to stay with the facts, with the aim to somewhat elucidate "what leads to what". Studying classic and recent review articles, reference lists of published letters and asking experts in the field, a scheme and routes are concluded and presented with arguements and selected representing literature.

Conclusions: Both hypo- and hyper-thyroidism constitute risk factor to preeclampsia. Additionaly, transient hypothyroidism is induced in mother and in utero, for teleological reasons, subject to regulation by an intact hypothalamus-pituitary-thyroid axis. The clinician should expect raised, normal or slightly pathological TSH up to a time relatively close to delivery. Counteracting influences lead to incalculable alterations in T_3, T_4, FT_3 and FT_4 and though a most common pattern coincides with lower FT_3 and raised FT_4, more diversity is allowed.

---

[*] e-mail: ixthioso@gmail.com.

TSH in Preeclampsia

In [2], it is stated that maternal T_3, T_4, FT_3, FT_4 are found decreased, whereas TSH is found increased (2.96 ± 1.07 vs 1.55 ± 0.89) when gestation age > 32 wks.

Similarly, in [3], both T_3 and T_4 are reported to significantly diminish in preeclampsia at third trimester and TSH is significantly increased in comparison to normal pregnancy (3.80 ± 0.53 vs 2.30 ± 0.24).

Following the same pattern, in [4], 3rd trimester TSH of preeclamptics is reported at 4.60 ± 3.64 vs 2.50 ± 2.01 in normal pregnancy.

Further in [5], TSH of control group is reported at 2.00 ± 1.18, of mild preeclamptics at 3.42 ± 1.61 and of severe preeclamptics at 5.63 ± 2.37, indicating increase with severity.

Also in [6], it is reported that patients with mild and severe preeclampsia showed significantly increased TSH, along diminished FT_3, FT_4 in severe cases compared to healthy controls. Increase in TSH in preeclampsia is also concluded in the nested case-control study of [1].

Any inverse relation between TSH and FT_3 or/and T_3 alone, indicates that the hypothalamus-pituitary-thyroid feedback loop is functional and intact finely ; and the inverse relation between FT_4 or/and T_4 and TSH, whenever observed, further advocates in favour.

The elevation of TSH in preeclampsia is a common finding in all the studies which have reported an alteration in TSH level. In fact, decreases in T_3, FT_3 are most commonly observed, alterations of T_4, FT_4 are reported bidirectional, but the rise in TSH is the feature highly prominent in the vast majority of these studies. The increase in TSH per se indicates at least subclinical or tissue -if not overt- hypothyroidism, in terms of increased demand, at least, in FT_3, irrespective of the amount of circulating serum FT_3 which may, indeed, be lower than normal or may be even unaltered [6].

The magnitude of the importance of TSH levels, hence the need for its fine regulation, in the pathology in question can be extracted from [7] according to which, interestingly, the frequency of gestational hypertension was lower in the women with subclinical hyperthyroidism (6.0 vs 8.8 %) in a statistically significant way ($p = 0.04$) and severe preeclampsia was somewhat more infrequent in the women with subclinical hyperthyroidism (3,5 vs 5.3 %) though in a statistically insignificant way ($p = 0.09$).

Further, in normal pregnancy and during 1st trimester, TSH falls due to a limited negative feedback by β-hCG which, as known, bears TSH activity. Correspondingly, the rise of TSH during preeclampsia, in which β-hCG is known to be elevated, further advocates strongly in favour of a maintained feedback "hypothalamus-pituitary-thyroid gland" loop, which can even surpass the normally expected small negative feedback upon TSH activity.

In [8] apart that the levels of TSH (and of Prl) in preeclamptic women during the 31st-35th week were found raised, it is reported that the responses to stimulation with TRH were the same between preeclamptics and the control group. Also, of note, cytokines IL-1β, IL-6, TNF-α and IFN-γ -which are found raised in preeclamspia ([9])- are known for inhibiting production of TSH (and TRH) ([10], [11]), further demonstrating the robustness in hypothalamus-pituitary-thyroid function is maintained.

One cannot omit to refer to the observation made in [8] that beyond 31-35 wks the differences between normotensive and preeclamptic individuals in TSH and Prl dissapeared. This can be conceded either by overdilution (which is inconsistent to preeclampsia, especially in severe cases), by establishment of reversible secondary hypothyroidism (which contradicts to the intact stimulation by TRH reported ibidem) or by resolution to a more euthyroid state, its teleological dimension being to provide for the changes around delivery. It is of interest to explore the behaviour of more close-to-delivery TSH levels as the said prospect could account partially for the discrepancies in literature concerning TSH levels, since different times of determination along gestation would entail different quantifications.

This concept can be further normalized by the following concessions: the small apparent reversion to lower values may occur at a time after 35th wk -not before- ; up to that time the demands in thyroid functions are increasing hence the elevated values of TSH but from that time on thyroid functions are partially satisfied at least at a local tissue level, either by decreased degradation of FT_3, FT_4 (probable route), by compensatory rise in deiodination, by increase in uptake from tissues (both occuring only very late, if so) or even by resolution to true euthyroidism.

Of note, elevated levels of TSH, when reported, still fall within or exceed by a few units the normal range determined by each study (usually based on normontensive pregnant women or even healthy non-pregnant women). This advocates further in favour of a finely maintained hypothalamus-pituitary-thyroid axis.

No elevation in TSH, this time not at late gestation age or versus control group, but in severe cases in compare to mild, was observed in [12] where TSH of uncomplicated pregnant women is reported at 1.02 ± 0.65 µIU/ml, of mild preeclamptic women at 1.47 ± 0.60 and of severe preeclamptic women at 1.32 ± 1.13, at 3rd trimester. The absence of marked elevation in TSH in severe vs mild cases contradicts to the mentioned assumed dependency from severity. Assuming there was no difference in the time between samplings in mild and samplings in severe cases (as more late in severe cases when, perhaps, normalisation has emerged), which is the most likely prospect, and given that no reversion to euthyroidism is expected in severe cases prior to mild cases, it can be said that the absence of elevation in TSH is due to decreased degradation of free thyroid hormones and loss of carrying proteins, both acting increasingly upon the means of the free thyroid hormones (of note, ibidem, FT_4 is reported increased in severe cases vs control values and even vs values of mild cases, while FT_3 is reported practically unaltered in severe cases vs control values). By this, even if reversion to euthyroidism has not emerged, TSH levels in severe cases can remain practically unaltered in compare to TSH levels in mild cases exactly by virtue of the severe pathology.

To the extreme side, in [13], no alteration of TSH even between severe preeclamptics and normotensive pregnant women is reported. Findings of this category obliges to a concesion either that in some instances there is no thyroid pertubations in preeclampsia (unlikely as this may stand, especially in severe cases) ; either that blood samplings occured too early along gestation or too close to delivery ; or that the study group included hyperthyroid preeclamptic individuals. By the inclusion of a subpopulation of (unreckognized or otherwise, at least subclinical) hyperthyroid pregnant women statistical process would bring about mean TSH values in preeclamptic groups more negatively biased toward means of control groups. This - assumed- inclusion has not to do with erroneous application of exclusion criteria in the control groups (which application is the usual pitfall in all studies) but involves the study group. The assumed inclusion of subpopulations with opposite pathophysiologies in the study group (if one conceeds that both hypo- and hyper-thyroid individuals are prone to development of preeclampsia, as also commented in the present) entails the need for studies where providence for patients with metabolic diseases known to affect thyroid function in the study group has been made.

Even so, why each research group reports one consistent pattern (f.e., only elevation or only lowering) remains open to debate.

Of note, two studies where such providence is clearly reported to have been made, conform with the majority according to which TSH is found increased in all cases ([4]) and in severe cases more than in the mild ones ([5]).

Presumably, reports for diminished TSH vs control are reserved only for hyperthyroid pregnant women as declared, f.e., in [14].

The additional factors contributing to the quantitative alterations of the thyroid hormones and accounting for the discrepancies in literature stand in the individuality (different genetical predispositions, pre-existing pathology such as diabetes mellitus, obesity, hypertension ; time between pregnancies, singleton or multiple pregnancy ([15]), the different pathophysiology between subpopulations of preeclamptics (apart between preeclamptics and HELLP patients) as now arises ([16]).

FT$_3$/FT$_4$ in Preeclampsia

In [6], where providence for classification according severity was made, it is reported that 3rd trimester patients with mild preeclampsia showed significantly increased FT$_4$ and TSH compared to healthy controls, while in severe cases, TSH was higher, with FT$_3$ as well as FT$_4$ be significantly lower than in controls. Other tests returned non significant differences.

Increased FT$_4$ and unaltered FT$_3$ in mild cases, could be explained by incomplete deiodination of FT$_4$ to FT$_3$ along downregulated degradation of FT$_3$ (f.e., decreased sulfation, glycuronidation, etc). The loss of TBG, of TBPA and of Alb, by virtue of proteinuria (due to kidney malfunction) and capillary leakage - hallmarks of preeclampsia - ; as well by virtue of faulty oestrogen production (due to placental dysfunction) is expected to further elevate FT$_4$ at this stage (for TBG in alone in preeclampsia, see f.e. [17] as well as [18]. However, loss of carrying proteins per se must not be the primary reason for elevated FT$_4$ because they would also have led to elevation of FT$_3$ were they intense enough, especially given that FT$_3$ is more influenced by protein changes than FT$_4$ ([19], [20]) and especially with FT$_3$ under decreased degradation. Decreased uptake of FT$_4$ from the tissues, even without reduction in deiodination, would have led, at some point, eventually to decreased FT$_3$ which contradicts to the report. β-hCG ([21]) which is found raised in preeclampsia, increased metabolic demands in this pathology ([22]) as well as mental stress could account for the rise in FT$_4$, but, for this, they would also have pushed towards elevation in FT$_3$. For the same reason, any theoretical increase of iodine uptake, in thyroid gland excretion, any raised activity of TPO, or of the symporter I$^-$/Na$^+$ can also be excluded as reasons for the elevation in FT$_4$. Decreased degradation of FT$_4$ without sufficient deiodination would have also led to increased FT$_3$. Since, however, any reason of decreased degradation of FT$_3$ is expected to also lead likely to decreased degradation of FT$_4$ and, further, rather simultaneously, since the catabolic routes are common (by the liver microsome enzymes) it is expected to coexist. Thus, a decreasing effect must be exerted upon the catabolism of both free hormones.

In each case, conclusively, incomplete deiodination should be considered to underlie.

Irrespective of the latter concerns, difference in FT_3 and FT_4 levels between severe and mild cases correspondingly, talks in favour of a gradual influence upon FT_3, FT_4, T_3 and T_4. Otherwise put, dependency of FT_3, FT_4, T_3, T_4 levels from severity must be assumed. This dependency of FT_3, FT_4 from severity is the likely cause of the disconcordance pertaining to totally conflicting observations when it comes to levels of thyroid hormones in preeclampsia – mainly in levels of FT_4, T_4. Namely, the degree of functionality remaining in liver and kidney of each individual can be the source of contradiction between the observations.

Further, the perturbation in deiodinases seems to occur almost in parallel and relatively simultaneously to the downregulation in degradation of FT_3. At this point, it is not possible to ascertain the priority between reduced deiodination, donwnregulation of any thyroid hormone, and urine loss of carrying proteins – one can only speculate upon their intensities comparatively and that all these events derive from liver malfunction.

Collectively, in mild cases, FT_4 can rise primarily by virtue of reduced deiodination and further due to reduced carrying proteins and decreased degradation. In mild cases, decreased production of FT_3 through deiodination appears to be occasionally potent to -more or less- counterbalance its reduced degradation and the loss of carrying proteins.

In severe cases, FT_4 can start to decline, probably because the raising influence of diminished carrying proteins and of reduced deiodination cannot anymore surpass another dominating event which manifests itself – decreased production, which remains the only reason to account for the said decrease. In severe cases, FT_3 also starts to decline -if has not already started, that is- because its reduced production outweighs the loss of carrying proteins and its reduced degradation.

The mentioned decreased production of FT_4 as severity marches on could be attributed to decreased tissue uptake of iodine which is reflected in iodine excreted in urine judging from [13] where it was shown that urine iodine of preeclamptic patients is found significantly and widely decreased (4.25 ± 2.7 vs 20.89 ± 6.4 mg/dl). The said diminishment led to the point that urine iodine is suggested as useful marker in preeclampsia and to the point iodine supplementation is suggested. No difference was found between urine T_4, FT_4 and TSH but there was a significant rise in urine T_3 and FT_3. This differential loss of T_3 and FT_3 through urine offers another route to how serum T_4 and FT_4 of preeclamptic women are occasionally found unaltered or even raised, whereas T_3, FT_3 are almost always found diminished, apart, or rather coexisting with, insufficient deiodination and reduced TBPA, Alb.

In addition to the participation of iodine, in [23], the lack in (serum protein-bound) iodine (PBI), indicated by the previous study, is confirmed, along a rise in umbilical cord blood PBI, in comparison to normal pregnant women, against prompting to the prospect of iodine supplementation. The contribution of initial insufficiency in iodine and the reported rise in cord blood PBI remain to be examined further.

Increased uptake from tissues (f.e., upregulation of FT_3 nuclear or/and cytoplasmic membrane receptors) is not to be the mechanism for the diminishment of FT_3, because increased uptake would have led to increased TSH as observed but, eventually, TSH level would have normalized upon fulfillment of maternal tissue energy demands. However, given the highly maintained TSH level, increased uptake does not appear to be the case. Additionally, raised demands from maternal tissues may be consisted with normal pregnancy but appear less consistent with preeclamptic pregnancy where the insufficiency of spiral arteries to provide nutrients to the placenta is now established and an increased metabolism of maternal tissues obviously would have worsen things.

Thus, the conclusion drawn regarding FT_3 and FT_4 stands in that, most likely, diminished deiodination of FT_4 to FT_3 occurs due to perturbation in liver function. After all, most of FT_3 is produced peripherally (70-90 % in terms of T_3) and since kidney and liver specifically account for the majority of circulating serum FT_3 it can be said, a priori, that what occurs as prerequisite is, indeed, impairment of peripheral metabolism in kidney and liver.

This pattern could as well fit from older to newer studies. For example, in the older study ([24]), where T_3 significant fall is reported for preeclamptic patients, as well as significant fall in FT_3 for 3 out of 9 and significant rise in T_4 and FT_4 in comparison to normal pregnancies. Both rise in FT_4, T_4 and fall in FT_3, T_3 could be reasoned by insufficient deiodination of FT_4 to FT_3. Similarly, in [2], diminishment is reported for both maternal FT_3, FT_4 and in [25] only for maternal FT_4 of preeclamptics vs controls.

A more complete concept can be drawn by the newer study [26], where it is stated that though lower at admission in preeclamptic women, spontaneous normalization occured during pregnancy (and up to three months post partum) so that at their third trimester 80 preeclamptic women did not have significantly different thyroid hormones compared to 10 normotensive women. Focusing on FT_3, FT_4 these findings can be indicative of counterbalance between decreased degradation and increased loss of (thyroid hormone carrying) proteins on one side; and decreased production and decreased deiodination on the other side, along gestation age.

That there are counteracting effects and that their counterbalance regulates the levels of thyroid hormones is a valid prospect. However, the concept of counterbalance to the extent exact euthyroid levels are achieved while thyroid pertubation persists is complex enough to make aside for a more simple one, i.e. the one reserved, in the present, for TSH: a time point no sooner than late in 3rd trimester occurs where thyroid function reverses somewhat to normality, at least at tissue level, with the reversion likely resulting to pure euthyroidism rather than the net outcome of opposite thyroid dysfunctions and with the rest thyroid hormones to follow after TSH towards normalisation.

In [12], it is reported that, in mild cases, FT_3 and FT_4 are significanlty decreased vs corresponding control values while, in severe cases, FT_3 is approx. unaltered and FT_4 is practically bordeline increased vs corresponding control values, at 3rd trimester. This is not a case where re-establishment of euthyroidism can be assumed as TSH is reported elevated vs control values. In here, decreased degradation-loss in carrying proteins in severe cases seem to effect more upon FT_3, FT_4 rather than decreased deiodination-decreased secretion.

Similarly, in [4] no difference in FT_3, FT_4 and rise in TSH of preeclamptics vs normotensive individuals was observed. Due to the elevated TSH this can be seen as a case of counteracting infuences. Another reason for the absence of alteration is given within the same study, namely that it could be due to the fact the blood sample was taken, even though at the 3rd trimester, just at the time of the diagnosis of preeclampsia.

Similar comments are reserved for [13] where no alteration in FT_4 and TSH, but increased FT_3 were observed. Commenced normalisation could provide for restoration of TSH and FT_4 and even some counterbalancing transient increase in FT_3.

If so, one side observation is that increasing or decreasing influences upon thyroid hormones in the pathology in question seem to affect with different intension in each individual -or subpopulations of individuals- so that one cannot oppine accurately if in severe cases it is decreased degradation which always prevails, or if in mild cases loss of carrying proteins always prevails or always falls short, and so on.

In a study from Jordan ([27]) no alteration in FT$_3$, FT$_4$ and TSH was observed in preeclamptics vs normotensive controls in any group of the three with gestation ages of 30-33, 34-37 and 38-41 wks. In this case, one can attribute the total absence of alteration only to different reasons in each sub-group. Namely, that it was too early for the pathology to affect thyroid function in the group of more early gestation age, that normalisation had started to occur in the group of latest gestation age and that the mentioned counteracting influences led to statistically compromised differences in the group of middle gestation age. Besides this complex prospect, extreme mildness in these particular study groups could have also been conceded, however, the characterization was based on objective criteria (urinary albumin, uric acid, systolic-diastolic BP) as, of course, expected. The fact that there was additional stratification according gestational age limits the sources of error even further. An option which could appear explanatory is that the study included hypo- and hyper-thyroid patients leading to maintenance of the mean values, namely the population could be biased ; again not likely to bring about maintainance of all values. The final overall conclusion drawn in the aforesaid study that "thyroid function is not altered in severe preeclampsia" could be taken as valid, with the meaning of that the "hypothalamus-pituitary-thyroid" loop remains intact, as extracted from other studies.

T$_3$/T$_4$ in Preeclampsia

The scope developed for the free hormones, which are the biological active forms, can fit in the total hormones as well.

A difference with the free fractions stands in that the loss of carrying proteins, this time, acts, additionally, to decrease the total fractions. A note based on this observation is that one would expect more frequent observations reporting raised serum FT$_4$ than raised T$_4$ in preeclamptic individuals. Also, similar to the free hormones, at this point, one cannot assume the priority between reduce in deiodination, loss of carrying proteins, assumed decrease in production of thyroid hormones and assumed decreased catabolism since all events equally derive their origin from kidney malfunction.

In [3], both T$_3$ and T$_4$ are reported to significantly diminish in preeclampsia at third trimester.

Similarly, in [2], it is stated that maternal T$_3$, T$_4$ are found decreased. It can be conceded, similarly to the free hormones, that these observations correspond to a stage when T$_4$ has eventually succumbed to declination due to reduced production which outweighs its reduced deiodination and assumed decreased degradation, whereas T$_3$ also decreases due to the aforesaid diminished deiodination.

Another instance is "[24], where T$_3$ significant fall is reported for preeclamptic patients, as well as significant rise in T$_4$ in comparison to normal pregnancies. Again, the pattern described can fit in, i.e. the concession is the occurrence of a mild/early stage when T$_4$ is still being raised by virtue of reduced deiodination while T$_3$ has started to decline even well before T$_4$.

Also, in [28], reduction in T$_3$ levels, rise in T$_4$ and rise in thyroxin binding capacity in pregnant women with late toxicosis is reported.

As assumed for the free fractions, strict order of intension cannot be assigned to the counteracting influences. Namely, T$_3$ may start to decline before the rise of FT$_4$ and T$_4$ in mild cases. This priority of T_3 against T_4 -when observed- can be reasoned by the fact that 1) T_3 is below T_4 in their catabolic conversion scale and thus is expected to alter more rapidly and by the fact that 2) T_3 exhibits much lower half-life in plasma (approx. 24-36 hours) in comparison to T_4 (5-7 days), which is due to less avid (about 100 times concerning TBG) binding of T_3 to carrying proteins in compare to T_4 ([29], [30]).

In [17], it is reported that, at their third trimester, preeclamptic patients had slightly lower TBG and significantly lower T_4, which were significantly correlated. In addition, there was no correlation between T_4 and TBG, in normotensive women.

The notation that there was no correlation between T_4 and TBG, in normotensive women could be taken as indication that the diminishment of T_4 in preeclampsia cannot be attributed solely in the diminishment of TBG and, further more, the said diminishment of TBG is not a major diminishing cause, otherwise no observation about raised T_4 level would ever have been made.

Further Observations

In an effort to pronounce upon the said deiodination malfunction, it is expected to stand in quantitative or qualitative perturbation in type I 5′-deiodinase or/and a perturbation in type II 5′-deiodinase, based on that these two are responsible for the production of T_3.

Also, judging from the location of type I 5′-deiodinase, i.e. the liver and kidney, this appears to be the enzymatic system more likely to be influenced over type II 5′-deiodinase, an assumption advocated by that type I deiodinase provides T_3 to the systemic circulation while type II deiodinase provides T_3 also locally, at the critical tissue level and, thus, one would expect a perturbation of this enzyme somewhat less easily achieved.

However, based on the location of type III deiodinase (brain, placenta), a perturbation of this enzyme cannot be ruled out either, given the established adverse symptoms upon brain and upon, of course, the placenta in preeclamptic patients. On the other hand, type III deiodinase protects the foetal circulation from excessive concentrations of T_4 and T_3 ([31]) and ([32]) and lies in high levels, thus, teleologically, its regulation should be maximally protected.

Thus, collectively, 5′-deiodinase type I is the most probable malfunctioning one and type II and type III probably succumb to malfunctioning along the progress of the disease.

In the case of perturbation in type I 5′-deiodinase one would expect counterbalancing increase in type I 5-deiodination leading to increased rT_3 (given the reciprocal relation between T_3 and rT_3). However, type I 5-deiodinases are, probably also affected in preeclampsia since they lie at the same location as type I 5′-deiodinases. By this assumption alone, serum rT_3 is expected to decrease along serum T_3.

In the case of perturbation in type II deiodinase one would expect a slight systemic and local rise in rT_3 at the corresponding tissues due to the lack of 5′-deiodination.

In the case of perturbation in type III deiodinase one would expect a local decrease of rT_3 at the corresponding tissues.

On the other hand, coming to the clearance and degradation of rT_3, these latter parameters are expected to decrease, given the established liver and kidney malfunctioning. By this assumption alone, serum rT_3 would tend to increase. Thus, the final behaviour of serum rT_3 subjects to these opposing influences.

Table 1. Lifestyle, environmental and pathological events leading to decreased T$_3$ and rT$_3$, i.e. alteration in peripheral conversion of thyroid hormones

Aging	(risk factor for preeclampsia)	[34]
Free radical load	(symptom of preeclampsia)	[35]
Insulin-dependent diabetes mellitus	(risk factor for preeclampsia)	[36]
Liver disease	(symptom of preeclamsia)	[37]
Kidney disease	(symptom of preeclamsia)	[38], [39]

Table 2. Deiodinases included in the peripheral catabolism of T$_4$ in humans ([40], [41], [43], [44])

Isoform	Location	Activity	Result
Type I deiodinase	Liver, kidney, thyroid, pituitary	5′-deiodination	systemic ↑T$_3$, ↓rT$_3$
		5-deiodination	systemic ↑rT$_3$,↓T$_3$
Type II deiodinase	Brain, pituitary, BAT, skeletal muscle, heart, thyroid	5′-deiodination	systemic+local ↑T$_3$, ↓rT$_3$
Type III deiodinase	Brain, placenta, uterus, skin, fetus	5-deiodination	local ↑rT$_3$, ↓T$_3$

Of note, under normal conditions 45-50 % of T$_4$ is turned into rT$_3$, a percentage subject to variation ([33]).

Even though there are opposing factors and assumptions which cannot evade of individualization, it is logical to expect at least a phase of somewhat raised rT$_3$. Though there are limited studies correlating preeclampsia and serum rT$_3$, interestingly some causes leading to decreased T$_3$/increased rT$_3$ are currently contemplated risk factors for preeclampsia or accompanying symptoms of preeclampsia (Table 1)

When it comes to the levels of thyroid hormones, another factor to be considered is hypovolaemia (haemoconcentration) which could be a finding consistent with prealampsia, in contrast to haemodilution in normal pregnancy. Though it remains to be examined the extent to which it modifies the relative levels of thyroid hormones, since hypovolaemia tends to increase the concentrations of all thyroid hormones, one can assume that the determined decreases in thyroid hormones, whenever observed, imply in fact even greater reductions in the absolute amounts of the corresponding hormones.

Thyroid Function as a Link between Preeclampsia and Eythyroid Sick Syndrome (ESS)

The fact that in most studies FT$_3$, T$_3$ of preeclamptic patients is found diminished whereas FT$_4$, T$_4$ are reported diminished, normal or elevated, brings in mind the variety in T$_4$ occurring in the ESS or low T$_3$ syndrome. Other common features are elevation of cortisol (however, in preeclampsia it is the placental cortisol which is elevated whereas serum cortisol remains unchanged comparing to normal pregnancy ([45]), decrease in serum albumin (for ESS: [34]), compromised selenium status (for ESS: [46] [47]), elevation of IL-6, TNF-α, IFN-α. [for preeclampsia: [9]); for ESS: [48]). However, preeclampsia declines from this syndrome in that in this syndrome TSH is either normal or slightly suppressed (when raised, underlying primary hypothyroidism preexists), following the general pattern according to which virtually in any case of increased secretion of endogenous cortisol, lowering of TSH occurs ([49], [50] [51], [52], whereas in preeclampsia TSH is much more commonly found elevated. In addition to maintenance of TSH levels, no low thyroid symptoms, neither remarkable findings on examination of thyroid gland occur in ESS, leading to the conclusion that it is probably a protective metabolic effect which lowers tissue requirements in the face of systemic illness (systemic illnesses are the usual cause of ESS) - however the prospect of tissue hypothyroidism cannot be ruled out. The same protective feature in the induction may fit as well in preeclampsia. Of note again, however, in preeclampsia TSH rises, which indicates true established hypothyroidism, in contrast to ESS.

To remind that ESS, at present is attributed most probably to decreased peripheral conversion of T_4 to T_3, which explains the preferential decrease first in T_3. This specific feature also fits in preeclampsia, where kidney, liver, placenta and also brain functions are highly influenced. To bear in mind that type II deiodinase is responsible for $\approx$ 70 % of T_4 deiodination in normal individuals, while in systemic illness type I deiodinase becomes the predominant deiodinase. The decrease in T_4 occurring in ESS and preeclampsia can be accommodated by common routes as well.

In ESS, also, rT_3 is found elevated due to the preferential inhibition of 5′-deiodinases leading to counterbalancing increase in 5-deiodination or/and due to reduced clearance of rT_3 (the latter occurs in non-thyroidal illnesses, NTI) ([53]). Also, in relation to preeclampsia and ESS, IL-1β, IL-6, TNF-α and IFN-γ are known to decrease type I deiodinase mRNA in vitro in rat thyroid cells and, interestingly these cytokines are raised both in ESS ([54]) and preeclampsia.

Table 3. Listing reasons leading to decreased circulating T_3

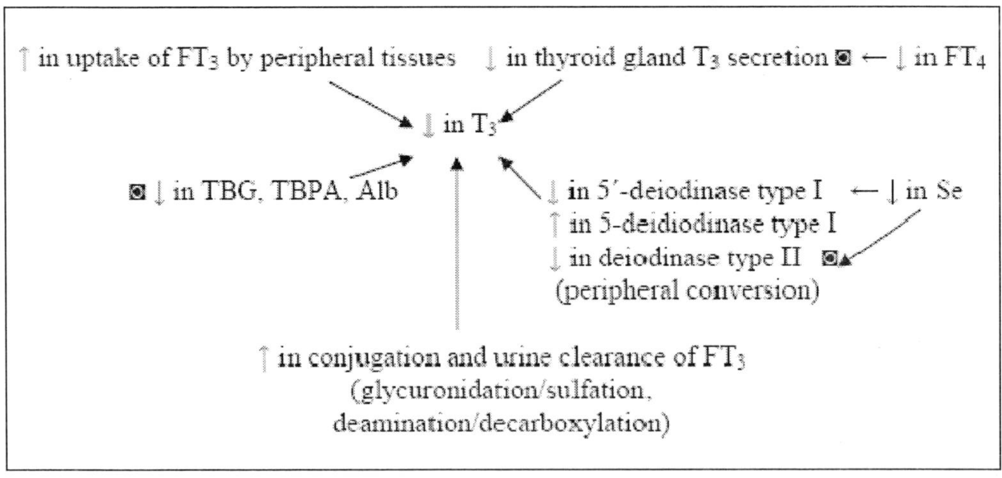

Table 4. Listing reasons leading to decreased circulating T$_4$

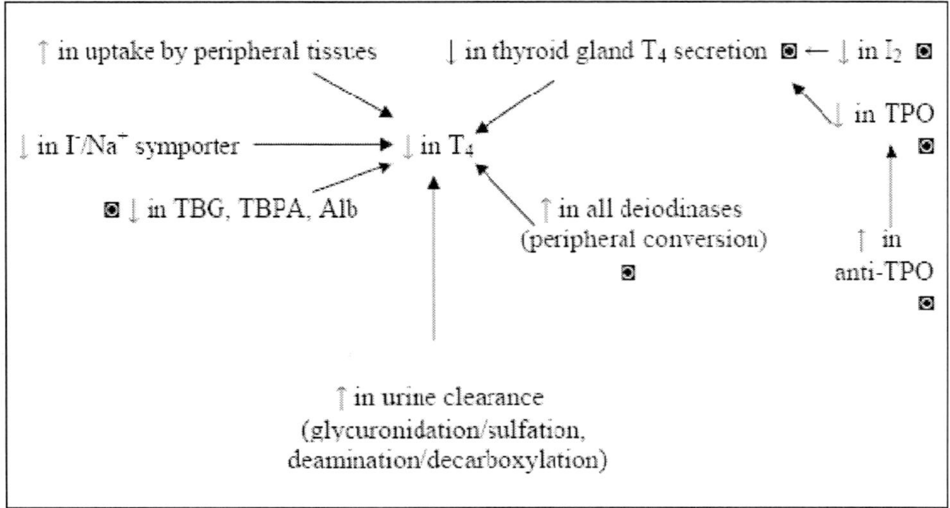

In NTIs, a most common cause of ESS, TBG is found almost unaltered, unless kidney function is pertubated due to the underlying disease. Thus, it has been speculated that the diminishment mostly in T$_4$ cannot be attributed to the quantitative decrease of TBG, but either rather to its impaired binding to T$_4$, to a coexisting binding-inhibitor or to rapid protease cleavage ([55], [56]).

Regardless of the validity of these prospects which are arguable, in preeclampsia there is no need to consider additional reasons for the lowering of T$_3$, T$_4$ such as the above, since proteinuria and capillary leakage consist hallmarks of the disease and can account equally well on their own both for the decrease in TBG and T$_3$, T$_4$. Where ◙ denotes a probable route.

From probable routes, the predominant ones stand in perturbation in 5′-deiodinase I and the fall in carrying proteins.

Conjugation of outer (phenolic) ring with glycuronides (by UDP-glucoronosyl-transferases) ([57]) or sulphate (by hepatic phase II sulfotransferases) leads to biliary and urinary excretion. Since the said conjugation occurs primarily by the said enzymes in liver microsomes and to a lesser extent in kidney, which are primarily affected in preeclampsia, the conjugation path is expected diminished. Oxidative deamination or/and decarboxylation of the alanine side-chain of the inner (tyrosyl) ring, occurring at liver microsomes ([58], [59]) resulting in the formation of the so-called acetic acid-analogue (tetraiodothyroacetic acid, metabolically active) is also expected decreased.

Hypo-, Hyper-Thyroidism and Preeclampsia: Defining a Cause – Result Axis

In [1], diminishment in FT$_3$ is reported, once again the elevation in TSH is brought up, along correlation of raised TSH with, soluble fms-like tyrosine kinase 1. The study concludes to 1) the establishment of subclinical hypothyroidism (24 % of 141 cases, adjusted odds ratio 2.2 with 95 % conf. interval 1.1 to 4.4) in mothers, and 2) the possibility of predisposition to hypothyroidism in later years, in a horizon of 20 yrs. This contradicts to the accepted reassuring spontaneous resolution of eclampsia after birth of the placenta (reassuring unless end stage renal failure or cerebrovascular haemorrhage have been induced meanwhile). Further, (pre)eclampsia is associated with increased risk or cardiovascular disease in a horizon of many years after births. A connecting mechanism between (pre)eclampsia and long-term cardiovascular burden might stand in subclinical hypothyroidism. Of note, only 12 % of 241 women with a history or (pre)eclampsia had raised TSH, thus follow-up studies are surely needed to confirm or reject the prospect of subclinical hypothyroidism being more frequent after (pre)eclampsia. Apart the apparent need for further studies, this particular study has been criticized strongly [The site of the national health service of the Department of health, http://www.nhs.uk/news/2009/11November/Pages/pre-eclampsia-pregnant-thyroid-gland-problem.aspx]. Specifically, the pitfalls stand in the following: i) principally, it is unclear whether blood test results that indicated an underactive thyroid were associated with any signs or symptoms of disease and if any hypothyroidism seen was severe enough to require treatment ; ii) it is not known whether thyroid function went back to normal following birth, whether it persisted and for how long ; iii) the first case control study was not originally designed to investigate the association between pre-eclampsia and thyroid function. It was a trial to investigate the use of calcium during pregnancy, and the women selected for the study may not be representative of pregnant women in general ; iv) the cohort study phase only measured thyroid function following birth. It is not known how this compared to pre-pregnancy levels ; v) it is not clear whether the health of the children in these studies was in any way affected.

For these, it is not possible to say whether pre-eclampsia contributes to underactive thyroid or if thyroid problems contribute to pre-eclampsia from this study.

The opinion of the authors of the present inclines towards the induction of hypothyroidism to the mother from preeclampsia and the teleological prospect of this induction can be the diminishment of pregnancy-raised but preeclampsia-impeded general maternal metabolic demands or/and the rise of very specific beneficial events other than net enegy balance to allow with priority for metabolic processes towards the infant. Most possibly, since pregnancy is a state of raised metabolic demands, the assumingly induced hypothyroidism stands far from the ideal physiology, however it may be the best it can be done under the circumstances once preeclampsia establishes itself in order a totally adverse outcome is avoided. However, it is felt that the mildness and the transient nature (unpublished observations) of the assumingly induced maternal hypothyroidism should also be intoned so as to add to a more complete concept.

Further concerning postpartum thyroid function, in [4], the observation is made that at no time during the postpartum follow up (6 wks) did any of the study women have symptoms and clinical signs of hypothyroidism. This finding does not necessarily contradict to the hint pertaining to induction of subclinical hypothyroidism made in [1], because the follow up period was rather small and, also, the reference concerned clinical signs and not subclinical biochemical findings.

In [60], it is reported that (pre)eclampsia has been observed in 16,7 % of subclinical hypothyroid patients and in 43,7 % of overt hypothyroid patients.

Subclinical hypothyroidism is most often caused by chronic autoimmune thyroditis, but also after damage to thyroid tissue through surgery, radiation, drugs. In this study, another rationale is presented, namely (pre)eclampsia-related vascular injury within the thyroid secondary to high concentration of the antiangiogenic factor Sflt-1.

In relation to this prospective, the vasodilatory molecule nitrix oxide, regulates secretion of thyroid hormones by modulating regional blood flow or activating follicular cells in the gland of animals ([61]), while its release has been found changed in rats with hypothyroidism ([62]). Complementary to the latter, immunohistochemical studies showed that progression of goiter in rat thyroids was accompanied by increased capillary endothelial cell growth ([63]).

Further adding to the prospect of antiogenetic factors and hypothyroidism is [2], where it is stated that a positive correlation between endothelin and TSH exists.

An even less obvious connection between subclinical hypothyroidism and preeclampsia is hinted in [64] according to which women with (overt or sublinical) hypothyroidism before pregnancy had higher incidence of gestational hyperthension compared to the general population. Ibidem, it is commented that low birth weights in case of hypothyroid mothers might be due to pre-term delivery of higher incidence, which is related to gestational hypertension. Similar to this less obvious setting, in [65] the frequency of pregnancy-related hypertension as well as of severe preeclampsia was reported similar between the study group and the control group but earlier delivery (as well as placental abruption and respiratory distress of infants) was more frequent in women with subclinical hypothyroidism.

Taken together, these observations, along the commonly reported rise in TSH, incline towards (even subclinical) hypothyroidism be a predisposing factor.

In [14], it is stated that significant higher incidence of preeclampsia was observed in women with hyperthyroidism (26.0 %) and hypothyroidism (26,8 %), preeclampsia diagnosed at third trimester, which inclines one to attribute roughly equal power of induction of preeclampsia from both pathophysiologies.

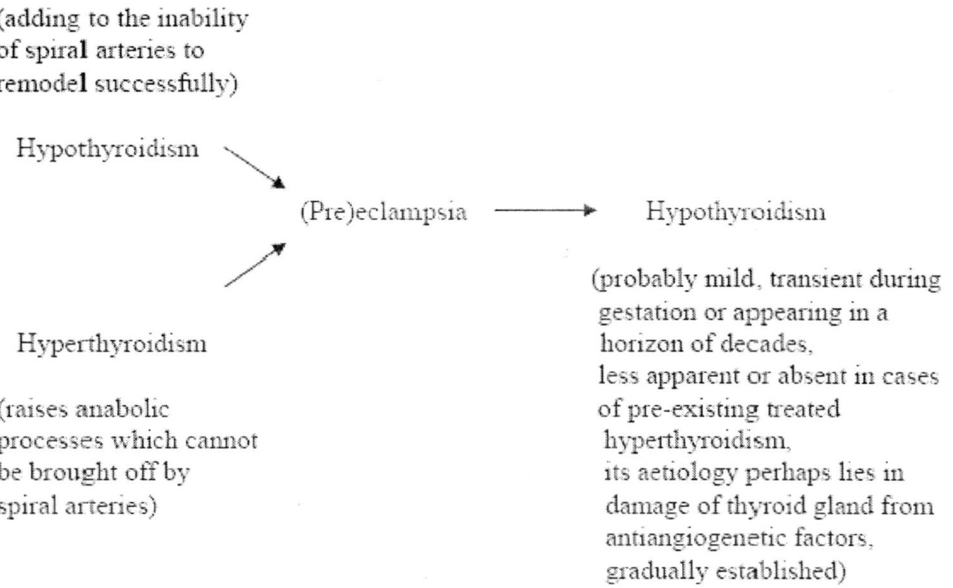

Figure 1. Thyroid and (pre)eclampsia in mother.

Commenting on hyperthyroidism, the sole fact that uncontrolled overt hyperthyroidism during pregnancy raises the risk of (pre)eclampsia $\approx$ 5-fold ([66]), leads strongly one to designate hyperthyroidism as a predisposing factor to (pre)eclampsia.

In [67], although the presence of eclampsia and marked hyperthyroidism is very rare, a case is reported which illustrates the importance of aggressive medical management against hyperthyroidism, demonstrating that hyperthyroidism is a causal risk factor for (pre)eclampsia.

On the other hand, in eythyroid patients with preeclampsia, the authors found that the T_4 and FT_4 was somewhat lower and TSH was a little higher in comparison to healthy pregnant women, advocating in favour of hypothyroidism also resulting from (pre)eclampsia.

Of note, ibidem, women with hyperthyroidism and (pre)eclamspia had T_4 and T_3 somewhat higher and TSH little lower than women in equal metabolic condition.

This single study does provide all data for placing hypo- and hyper- thyroidism along the "cause-result" axis and provides a prospect which we find rational. By utilizing findings from all other studies reported in here, a scheme is presented which unifies almost all drawn conclusions.

In partial support of this scheme one can also the roughly empirical, clear statement which is part of the guidelines of American Thyroid Association (Thyroid Disease and Pregnancy Chapter 2: Hyperthyroidism and pregnancy ; Chapter 3: Hypothyroidisim and pregnancy, 2005, www.ata.org) that both hypo- and hyper-thyroidism constitute risk factors for preeclampsia.

The fact that cytokines IL-1, IL-6, TNF-α, IFN-β are known to decrease the binding capacity of T_3 nuclear receptors ([68], [69]) and that they are raised in preeclampsia, talks in favour of hypothyroidism resulting from preeclampsia.

The fact known to all those familiar with clinical routine, namely that many antiepileptics and lithium (which had found place as symptomatic treatment against convulsions in preeclampsia) diminish T_3 and T_4 ([70], [71], [72], [73], [74]) advocates slightly in favour of that hypothyroidism results from preeclampsia and, at least, for neuro-protective reason. The anticonvulsant drug of choice used against preeclampsia at present, i.e. $MgSO_4$, can also serve as a connective link between thyroid function and preeclampsia.

It has long been known that serum Mg is elevated in (even subclinical) hypothyroidism and decreased in hyperthyroidism ([75], [76], [77]), that the erythrocyte Mg, as induced secondarily by the change in plasma concentrations, is also higher in hypothyroid and also lower in hyperthyroid ones ([78]), and that the excretion of magnesium is higher in hyperthyroid and lower in hypothyroid patients ([79]), with reversion to the corresponding levels of euthyroid patients after therapy of the thyroid disease.

Table 5. Some established effects of Mg at the cellular level

Growth/apoptosis	Increase in RBC deformability
Differentiation	[92], [93]
Inflammation	Blocking release of acetylcholine
Vasoconstriction/vasodilation	Blocking N-methyl-D-aspartame (NMDA)
[89]	glutamate receptors
Decrease in intra-erythrocytic Na^+	[95]
[90], [91]	

Altered Mg levels results from the perturbations in thyroid function (mostly in T_3 levels) ([80]) at least by means of altered glomerulal filtration rate ([78]) as well as by the activation of the rennin-aldosterone system by thyroid hormones ([81]). Some authors have reported no difference in plasma Mg of either hyperthyroid or hyperthyroid individuals (f.e., [82]), but these probably are attributed to the unavailability of atomic absorption spectrophotometry for determination of Mg levels or to the absence of radioimmunassay to determine thyroid hormone levels at that time (some 60 yrs ago) ([78]).

On the other hand, in preeclampsia, serum total and ionized Mg significantly lowers, in compare to normal pregnancy (during which Mg also lowers), with increasing gestation age as shown by most studies (f.e., [83], [84]), though some have shown no difference ([85]) reflecting perhaps differences in gestation age (however, even in the contradicting studies the intra-erythrocytic Mg^{2+}, when determined, still is reported lower) and while some have shown even elevated serum total ([86]) as well as ionized Mg ([87]).

Thus, in terms of opposing to preeclampsia-induced diminishment of magnesium, the presumed preeclampsia-induced hypothyroidism appears to act beneficially. Empirically, the significance of the effects exerted by Mg (Table 5) in preeclampsia and in various pathophysiologies is now well established, at the extent that even a "magnesium-ischaemia" hypothesis has been suggested ([88]) ; it is, thus, likely the regulation of the levels of Mg is placed in a high priority over other various metabolic processes in the pathophysiology in question. The induction of somewhat compensatory hypothyroidism could serve the cause.

One selected effect from the aforementioned which has been verified in preeclampsia stands in the increase in RBC deformability. $MgSO_4$ therapy in vivo and in vitro stands in reducing erythrocyte osmotic fragility in preeclamptics (according to the authors' interpretation by means of reaction with free radicals) ([96]), another beneficial effect of Mg. Given the elevation of Mg levels in hypothyroidism, the assumed induction of hypothyroidism in preeclampsia seems to exert an extra beneficial effect, minor as the contribution of fragile RBCs in the pathology may be. Admittedly, hypothyroidism per se has no effect on the fragility of RBCs of non-pregnant hypothyroid individuals ; in fact, it is hyperthyroidism which per se lowers the erythrocyte fragility in common hyperthyroid patients ([97], [98]) but one has to consider than the osmotic fragility of RBCs in preeclampsia is raised in compare to normal pregnancy (thus, even a little rise in serum Mg could stand beneficiary when it would not make such a benefit upon RBCs of normal fragility).

There are various parameters in the rheology during preclampsia and during thyroid pertubations which can be taken in mind in the effort to designate the "cause-result" priority.

In preeclampsia and in some cases of IUGR, RBC microcirculation and deformability are reduced ([92], [93]). Because secreting glands are adversely affected by red cells of poor deformability, it is not surpring thyroid malfunction should ensue.

Because whole blood viscosity is a function of plasma viscosity, haematocrit and red cell deformability (among others such as, f.e., IgM, lipoproteins and Alb content) ([94]), by means of reduced RBC deformability and by hemoconcentration (tending to elevate the Hct), whole blood viscosity is also raised in preeclampsia ([99], [100]) in contrast to normal pregnancy where whole blood viscosity decreases. However, hyperviscosity has not been consistently observed in all clinical studies ([101]). Changes in blood viscosity can be compensated readily in the normal circulation but not in the compromised, low-flow circulation, thus, blood viscosity –when present (with providence for the reported lack in alteration) - contributes to the phenotype of preeclampsia ([94]) since viscosity is a constituent of vascular resistance.

On the other hand, underactive thyroid function has also been correlated to raised whole blood viscosity ([103], [104]) even after adjustment at Hct 40%, with the viscosity not correlating to lipoprotein concentrations ([105]). The viscosity can be considered secondary to hypothyroidism ([102], [105]).

Though it is tempting for one to speculate that the ensuing raised viscosity in preeclampsia is mediated by hypothyroidism, the far more probable concept is that hyperviscosity in preeclampsia is secondary to increased vascular permeability and that underactive thyroid function contributes further to it. From the prespective of rheology, thus, hypothyroidism is expected to contribute in the symptoms when preexists, and to further burdens if it is induced. Othewise put, from the prespective of rheology, designating hypothyroidism as a risk factor is further strengthened ; also, the concept is brought forward that hypothyroidism, if teleologically induced, is a "less-than-ideal" state, with beneficial effects - perhaps the ones necessitated most by the organism- but also, with burdening effects.

Another metal which could provide knowledge upon the connection of preeclampsia and thyroid function stands in zinc. In preeclamptic women zinc is commonly found diminished in serum, in leukocytes ([106], [107]), in placenta ([107], [108]) in compare to normal parturients as well as to chronic hypertensive parturients, though no alteration has also been reported, less commonly ([109]). The cause of loss could, assumingly, stand in a loss of major transport proteins of zinc namely albumin and α-2 macroglobulin due to proteinuria or/and capillary leakage.

Further, as known (f.e., see [110]), zinc promotes thyroid function in humans, at least, by zinc finger-transcription factors which occur at thyroid hormone receptors and are necessary for the thyroid hormones to exert their effect after binding ([111]), with autoimmunity emphasized]. Mean plasma zinc concentration in hypothyroid patients is reported lower and urinary zinc excretion is reported increased in hyperthyroidism ([112]). Oral zinc supplementation has increased thyroid function in children with Down syndrome ([113]), in uremic patients under peritoneal dialysis and in disabled patients [as reviewed thoroughly in [114].

Obviously, the diminishment of zinc in preeclampsia may not suffice per se for a profound hypothyroid effect but it offers additional indication in favour of maternal as well as in utero hypothyroidism. The establishment of in utero hypothyroidism can be hinted by significant decreases in cord plasma zinc of preeclamptics ([115]) as well as by decreases in the placenta ([107], [108]).

The opinion of the authors of the present is that deficiency of zinc cannot hold any primary causal role in preeclampsia first because lack in zinc is a rare nutritional deficiency (this is why supplementation with zinc is not recommended during pregnancy on customary basis), second because the correlation between zinc levels and birth weight observed in preeclamptics is not observed in normal groups ([116]) thus the deficiency can be contemplated as result of the underlying pathology) and third because two implemented randomized studies involving supplementation of preeclamptics with zinc have provided poor outcomes ([106]), [117]) ; but, it is logical to assume that its deficiency contributes in lowering thyroid function.

Of note, the element of copper appears elevated in the serum ([118], [108]) and in the placenta of preeclamptics ([108]) and, as known, copper exerts a recognized lowering effect upon thyroid, indicating, again, an extra minor factor in favour of maternal and in utero hypothyroidism Lowered copper values have also been observed in preeclampsia in compare to normal pregnancy ([119]) which could be indicative rather of the minor significance of the element' s absolute amount and of the greater significance of the relative levels of the element (with respect, f.e., to its antagonist, zinc).

In [120], the contradicting observation is made that zinc levels rose in the particular study group of preeclamptics. However, even in this study, a higher ratio of plasma copper/zinc was found, still serving the assumption for suppression of thyroid function by means of this altered ratio.

A perhaps even less obvious connetion to thyroid stands in the element of iron. In hypothyroidism, lower amounts of serum iron occur (due to diminished secretion of hydrochloric acid and thus diminished absorption from the gastrointestinal tract), even to the extent sometimes the treatment against anaemia in hypothyroid individuals is in vain unless the underlyling hypothyroidism is treated. On the other hand, in preeclampsia serum iron is found raised ([121], [122], [123]). The said rise is attributed primarily to haemolysis, secondarily to hepatocellular damage, rather not at all to intracellular volume contraction ([124]), and perhaps ischaemic placenta also is a sourse ([121]). Thus, the said rise in serum iron cannot hold any primary causal role, but, logically, it is expected to contribute as a secondary burdening cause (due to overload of cells and, thus, raised oxidative stress) ([121]). One could see a beneficial effect of the assumed induced hypothyroidism in that it offsets further intake-induced rise in iron. That the organism tries to cope with the excess iron can be seen by the concomitant rise in ferritin which, as known, keeps iron dissolved and thus less toxic to cells is also raised in preeclampsia ([122], [123]) in contrast to normal pregnancy where ferritin decreases progressively reaching a nadir at the third trimester ([125]). Admittedly, the elevation of ferritin is not attributed to elevation of iron directly (after all, iron-induced elevation in ferritin would have been observed approx. in a 3 month-time span) but rather, primarily, to non specific release by virtue of hepatocellular damage ([122]). However, this lack in selective regulation of iron does not mean that this accompanying rise in ferritin cannot be an indirect, more quickly induced, mechanism protecting from serum iron overload, as, secondarily, induction from inflammatory cytokines (a route subject to regulation) also occurs.

Concerning the prospect of iron in preeclampsia, there have been studies which address with scepticism the issue of iron supplementation to uncomplicated well-nourished pregnant women with sufficient intake of necessary ingredients ([123], [125], 126]).

To be examined futher in preeclampsia remains the insufficiency observed for the trace of selenium which affects all three Se-dependent deiodinases ([127]). Lowered plasma selenium is reported even before the onset of preeclampsia ([128]), however, administration of selenium did not lead to clear results ([129]). Thus, at present, decrease in selenium is brought up as factor reducing thyroid function resulting from preeclampsia.

Anti-Thyroid Antibodies and Preeclampsia

In [130], anti-Tg and anti-TPO were found at 33,3 % of preeclamptic patients and at 14,5 % of the controls, the difference being statistically significant. Thus, this study rather favours the prospect of humoral immunity taking action against thyroid in preeclampsia. Of note, higher frequency of anti-thyroid Abs in early loss is known to those familiar with laboratory results.

Similarly in [26], it is stated that anti-TPO Abs were raised in 10 % of (preeclampsia, HELLP, or gestational hypertensive) patients 3 months post partum, indicative of a reminiscent motivated immune response.

However in [1], the absence of an autoimmune process (based on absence of anti-TPO Abs) in preeclampsia is brought up despite the establishment of subclinical hypothyroidism in 24 % of 141 cases.

In [66] it is stated that TSH-receptor Abs (thyroid binding inhibitory Igs) were not related to (pre)eclampsia, adding a small hint towards independence from humoral immunity as far as establishment-progress of (pre)eclampsia and thyroid function are concerned.

Preeclampsia and the Thyroid Function of the Infant

In [131], it is stated that mean cord blood TSH from 98 toxemic mothers did not differ significantly in comparison to healthy mothers (or to mothers with AIDS, diabetes mellitus, cardiovascular malfunctions) suggesting a protective mechanism -protecting at least up to a point- which allows for normal cord blood TSH levels in infants from mothers with toxaemia or other stressful diseases.

However, the said study contradicts to [132], in which significantly raised cord blood TSH was observed along positive correlation between TSH and preeclampsia, glucose intolerance and other maternal medical diseases (apart antepartum haemorrhage of unknown origin). Raised cord blood TSH was also observed in [133].

A third prospect concerning cord blood TSH in preeclamptics is seen in [26] where it is reported that the neonates from preeclamptic individuals had lower FT_4 and TSH, as determined through cord blood FT_4 and TSH measurements in comparison to infants of corresponding gestational age included in the control group.

The opinion of the authors of the present incline towards the third study, namely that preeclampsia affects the thyroid function of infants. The said influence can be seen as an (adverse and obligatory) adaptation of the embryo to mother' s inability to provide proper quantity of nutrients due to the known inadequate remodelling of spiral arteries which, at present, consists the earliest found underlying cause of (pre)eclampsia. This influence, f.e, could explain the transient hypothyroxinaemia seen in preterm infants after birth of preeclamptics ([133]) which is familiar to all those involved in obstetric routine. Developing foetal structures are exquisitely attuned to fine alterations in maternal fuel economy. For instance, the gestational interactions create a pattern of accelerated starvation whenever food is withheld ([134]).

The prospect of lower maternal energy output can be suspected by the said concomitant lowering also of cord TSH, namely by the image of in utero secondary hypothyroidism, which deprives the chance of positive feedback from hypophysis due to lowered T_4/T_3 (perhaps indirect indication of intact infant hypothalamic-pituitary-thyroid axis). Assumingly, once the energy demands of the infant can be satisfied after birth, reversion of suppressed thyroid function occurs relatively in brief time. This reversion of thyroid assays values and of any clinical signs would demand an intact hypothalamic-pituitary-thyroid axis which is exactly what is clearly suggested by the transient nature of hypothyroxinaemia in preterm infants. Since, however, both elevations and decreases of cord blood TSH have been observed, the said conception should allow for both prim- and sec-hypothyroidism, presumably depending from severity/progress of the underlying preeclampsia.

These conclusions exactly can be advocated by [135], where lowered cord blood FT_4, TSH and rT_3 are reported indicative of an underactive thyroid function before birth. Despite the lower values found, the conclusion reached ibidem is that preterm infants from preeclamptic mothers with placental insufficiency have intact hypothalamic-pituitary-thyroid axis which allows for subsequent observed normalization of thyroid assay values. Integrity of the hypothalamic-pituitary-thyroid axis is also concluded in [133] and also in [28]. Additionally, in [64] it is concluded that maternal hypothyroidism was not associated with poor foetal and neonatal outcomes. A yet–to–be–evaluated finding in [135] is that, on days 5 and 7, blood rT_3 was found raised in compare to preterm infants from mother without placental insufficiency which was taken as temporarily impairment of hepatic type I 5′-deiodinase. As commented in the present, this very impairment is the most probable one for the same type of maternal enzyme and it would be of interest to clarify the route of a similar enzymatic behaviour induced in the liver of the infant.

An effort to explore the route of induction can be found in [26] according to which cord blood FT_4 was found not to correlate with maternal FT_4 but with umbilical artery pH, indicative of prenatal acidosis. This provides a rough insight to the route by which preeclampsia influences thyroid function in embryos: the route must be different from direct influence from maternal thyroid function (f.e., due to diminished transplacental T_4 passage), but rather secondarily, owing to intrauterine stress, perhaps due to elevated metabolic acidosis or due to another preeclamsia-induced factor.

The findings referred in [25], i.e. that there is no significant difference in birth weights between preeclamptic women with low vs normal T_4 and T_3, offer additional support that the embryo is protected from the lack in maternal T_4. To this latter prospective, advocates [136] where umbilical FT_4 of very low birth weight infants was normal. Independency of umbilical FT_4 from maternal FT_4 is reported in [26].

According [136], umbilical FT_4 of very low birth weight infants was normal. The hypothesis of infant' s protection against the lack in maternal T_3 and T_4 especially through maintenance of FT_4, might explain the significant correlation between birth weight and infant' s T_3 and T_4, but not between birth weight and infant' s FT_3 and FT_4 in infants from (pre)eclamptic individuals.

In [2], the (positive) correlation between birth weights and T_3, T_4 and the (negative) correlation between birth weights on one hand and TSH and endothelin on the other hand is brought up for preeclamptic individuals, while not for normal individuals. The observation made ibidem, that the negative correlation between birth weights and endothelin in preeclamptics was found more significant than the mentioned correlations between birth weights and thyroid hormones, as well as the suggested absence of correlation between FT_3, T_3, FT_4, T_4, TSH and birth weight in the normal group, suggests that the $(F)T_3$, $(F)T_4$, TSH and birth weight in preeclampsia (at least in those cases studied, i.e. in those gestational age groups and of those severities) are independent covariables, i.e., they alter due to a third variable being their common cause, presumably endothelin. Other molecules highly pathognomonic for preeclampsia such as sFMS-like tyrosine kinase 1 or/and sEndoglin could also be the candidates for reducing birth weight.

Further birth weight seems independent, also, from TBG in preeclampsia, as suggested in [17]. It is reported that, at their third trimester, preeclamptic patients had slightly lower TBG and significantly lower T_4, which were significantly correlated. They also had significantly lower infant birth weight which was correlated to TBG. Further, there was no correlation between T_4 and TBG, as well as between TBG and birth weight, in normotensive women. The rationale of covariables which has been described in here for T_3, FT_3, T_4, FT_4 TSH and birth weight in preeclampsia, most likely suits the correlation between TBG and birth weight as well, a prospect enhanced by the absence of such correlation in normal pregnant women. In [4], the correlation between maternal FT_3, FT_4, TSH of preeclamptics and birth weights which has been reported by others ([18]), was not observed.

References

[1] "Pre-eclampsia, soluble fms-like tyrosine kinase 1, and the risk of reduced thyroid function: nested case-control and population based study", Levine RJ, Vatten LJ, Horowitz GL, Qian C, Romundstad PR, Yu KF, Hollenberg AN, Hellevik AI, Asvold BO, Karumanchi SA, *BMJ*, 2009; 339: b4336.

[2] "Correlation between maternal thyroid function tests and endothelin in preeclampsia-eclampsia", Başbuğ M, Aygen E, Tayyar M, Tutuş A, Kaya E, Oktem O, *Obstet. Gynecol.*, 1999; 94(4): 551-5.

[3] "Thyroid functions in pre-eclampsia and its correlation with maternal age, parity, severity of blood pressure and serum albumin", Khaliq F, Singhal U, Arshad Z, Hossain MM, *Indian J. Physiol. Pharmacol.*, 1999; 43(2): 193-8.

[4] "Maternal thyroid hormonal status in preeclampsia", Kumar Ashok, Ghosh BK, Murthy NS, *Indian Journal of Medical Sciences*, 2005; 59(2): 57-63.

[5] "Thyroid hormones in pregnancy and preeclampsia", Divya Sardana, Smiti Nanda, Simmi Kharb, *J. Turkish German Gynecol. Assoc.*, 2009; 10(3): 168-171.

[6] "Thyroid hormone alteration in pre-eclamptic women", Larijani B, Marsoosi V, Aghakhani S., Moradi A., Hashemipour S., *Gynecol. Endocrinol.*, 2004; 18 (2): 97-100.

[7] "Subclinical hyperthyroidism and pregnancy outcomes", Casey BM, Dashe JS, Wells CE, McIntire DD, Leveno KJ, Cunningham FG, *Obstet. Gynecol.*, 2006; 107: 337-41.

[8] "Serum prolactin and thyroid stimulating hormone levels following thyreotropin releasing hormone stimulation in preeclamptic patients", Kulseng-Hanssen S et al., *Acta Obstet. Gynecol. Scand.*, 1979; 58(3): 245-8.

[9] "Immunology of preeclampsia" Leif Matthiesen, Göran Berb, Jan Erneurudh, Christina Ekerfelt, Yvonne Jonsson, Surendra Sharma, Chem. Immunol. Allergy, Basel, Karger, 2005; 89: 49-61.

[10] "Different effects of continuous infusion of interleukin-1 and interleukin-6 on the hypothalamic-hypophyseal thyroid axis", GA van Haasteren, MJ van der Meer, AR Hermus, E Linkels, W Klootwijk, E Kaptein, H van Toor, CG Sweep, TJ Visser and WJ de Greef, *Endocrinology*, 1994; 135: 1336-1345.

[11] "Fundamentals of psychoneuroimmunology", Cai Song, Leonard B.E., 2000, John Wiley and Sons.

[12] "Thyroid function in preeclampsia", Maged El-Shamy, Hamed Youssef et al, Mansoura University Teaching Hospital, Mansoura, Egypt, 2005, reference letter.

[13] "Urine iodine levels in preeclamptic and normal pregnant women", Gulaboglu M, Borekci B, Delibas I, *Biol. Trace Elem. Res.*, 2009 Oct 29 (Epub ahead of print).

[14] "The effect of pre-eclampsia on thyroid gland function", Vojvodić Lj, Sulović V, Pervulov M, Milacić D, Terzić M, Srp Arh Celok Lek, 1993 Jan-Feb; 121(1-2): 4-7.

[15] "Risk factors for pre-eclampsia at antenatal booking: systematic review of controlled studies", Kirsten Duckitt, Deborah Harrington et al., *BMJ*, 2005; 330-565.

[16] "Severe preeclampsia with and without HELLP differ with regard to placental pathology", Marie-Therese Vinnars, Liliane C.D. Wijnaendts, Magnus Westgren, Annemieke C. Bolte, Nikos Papadogiannakis, Josefine Nasiell, *Hypertension,* 2008; 51: 1295.

[17] "Birth weight and thyroxine-binding globulin in pre-eclampsia", Lao TT, Chin RK, Panesar NS, Lam YM, *Gynecol. Obstet. Invest.*, 1989; 28(1): 8-10.

[18] "Maternal thyroid hormones and outcome of preeclamptic pregnancies", Lao TT, Chin RKH, Swaminathan R, Lam YM, *Br. J. Obstet. Gynaecol.*, 1990; 97: 71-4.

[19] "Effects of various contraceptives on laboratory parameters in diagnosis of thyroid gland function with special reference to the free hormones FT4 and FT3", Sorger D, Schenk S, Schneider G, *Z. Gesamte Inn. Med.*, 1992; 47(2): 58-64.

[20] "Increased serum concentration of free L-triidothyronine in patients treated with L-thyroxine", Schmitz R, Hotze A, Bongers H, Low A, Mahistedt J, Joseph K, Wolf F, *Nuclearmedizin*, 1983; 22(5): 251-4.

[21] Comparison of thyroid stimulators and thyroid hormones concentrations in the sera of pregnant women, Harada A, Hershman JM, Read AW, Braunstein GD, Dignam WJ, Derzko C, Friedman S, Jewelewicz R, Pekary AE, *J. Clin. Endocrinol. Metab.*, 1979; 48(5): 793-7.

[22] "Thyroid status in normal pregnancy", Burrow GH, *J. Clin. Endocrinol. Metab.*, 1990; 71: 274-5.

[23] "Iodine and magnesium levels in maternal and umbilical cord blood of preeclamptic and normal pregnant women", Borekci B, Gulaboglu M, Gul M, *Biol. Trace Elem. Res.*, 2009 summer; 129(1-3): 1-8, Epub 2008 Nov 26.

[24] "Total and free thyroxine and triiodothyronine in normal and complicated pregnancy", Osathanondh R, Tulchinsky D, Chopra IJ, *J. Clin. Endocrinol. Metab.*, 1976 Jan; 42 (1): 98-104.

[25] "Thyroid function in preeclampsia", Lao TT, Chin RK, Swaminatha R., *Br. J. Obstet. Gynaecol.*, 1988; 95: 880-3.

[26] "Transient hypothyroxinemia in severe hypertensive disorders of pregnancy", Buimer M., van Wassenaer A.G., Ganzevoort W., Wolf H., Bleker OP., Kok JH., *Obstet. Gynecol.*, 2005 Nov ; 106 (5 Pt 1): 973-9.

[27] "Severe pre-eclampsia and maternal thyroid function", Qublan HS., Al-Kaisi IJ., Hindawi IM., Hiasat MS., Awamleh I., Hamaideh AH., Abd-Alghani I., Sou'ub RM., Abu-Jassar H., Al-Maitah M., *J. Obstet. Gynaecol.*, 2003 May; 23(3): 244-6.

[28] "Functional study of the hypopheal-thyroid and adrenal systems in physiological pregnancy and late toxicosis", Tsil'mer KIa, Kallikorm AP, Paiu Alu, *Probl. Endokrinol.* (Mosk), 1985 Nov-Dec; 31(6): 16-21.

[29] "Review of medical physiology", 22nd ed., Ganong WF, 2005, The McGraw-Hill Companies.

[30] "Vertebrate endocrinology", 2nd ed., 1985, Norris D.O., Lea and Febiger.

[31] "Maternal and foetal thyroid function", Burrow GN, Fisher DA, Larsen PR, *N. Eng. J. Med.*, 1994; 331: 1072-1078.

[32] "Increased TBG during pregnancy and increased hormonal requirements", Glinoer D., *Thyroid*, 2004; 14: 479-480.

[33] "An assessment of daily production and significance of thyroidal secretion of 3,3',5' triiodithyronine (reverse T3) in man", *J. Clin. Invest.*, 1976; 58: 32-40.

[34] "Euthyroid sick syndrome, associated endocrine abnormalities, and outcome in elderly patients undergoing emergency operation", Girvent M, Maestro S, Hernandez R, et al.. *Surgery,* 1998; 123: 560-567.

[35] "The protective role of some antioxidants and scavengers on the free radicals-induced inhibition of the liver iodothyronine 5'-monodeiodinase activity and thiols content", Brzezinska-Slebodzinska E, Pietras B., *J. Physiol. Pharmacol.*, 1997; 48: 451-459.

[36] "Altered adrenal and thyroid function in children with insulin-dependent diabetes mellitus", Radetti G, Paganini C, Gentili L, et al., *Acta Diabetol.*, 1994; 31: 138-140.

[37] "Liver and thyroid gland. Physiopathologic and clinical relationships", Gallo V, Rabbia F, Petrino R, et al., *Recenti Prog. Med.*, 1990; 81: 351-355. [Article in Italian].

[38] "Selenium deficiency and thyroid function in acute renal failure", Makropoulos W, Heintz B, Stefanidis I., Ren. *Fail.*, 1997; 19: 129-136.

[39] "Effect of fushen decoction on thyroid hormone in chronic glomerulonephritis of both spleen and kidney-yang deficiency patients", Han MX, Wang YP, Chung Kuo Chung Hsi I Chieh Ho Tsa Chih, 1993; 13: 155-157, 133 [Article in Chinese].

[40] "Factors altering thyroid hormone metabolism", Robbins J., *Environ. Health Perspect*, 1981; 38: 65-70.

[41] "Flavonoid effects on transport, metabolism and action of thyroid hormones", Kohrle J, Spanka M, Irmscher K, Hesch RD., *Prog. Clin. Biol. Res.,* 1988; 280: 323-340.

[42] "Pathways of thyroid hormone metabolism", Visser TJ., *Acta Med. Austriaca*, 1996; 23: 10-16.

[43] "The deiodinase family of selenoproteins", St Germain DL, Galton VA., *Thyroid*, 1997; 7: 655-668.

[44] "Hormone-nuclear receptor interactions in health and disease. The iodothyronine deiodinases and 5'-deiodination", Beckett GJ, Arthur JR., *Baillieres Clin. Endocrinol. Metab.*, 1994; 8: 285-304.

[45] "Prematurity is related to high placental cortisol in preeclampsia", Aufdenblatten M, Baumann M, Raio L, Dick B, Frey BM, Schneider H, Surbek D, Hocher B, Mohaupt MG, *Pediatr. Res.*, 2009; 65(2): 198-202.

[46] "Plasma selenium concentrations in patients with euthyroid sick syndrome", Van Lente F, Daher R., *Clin. Chem.,* 1992; 38: 1885-1888.

[47] "Relations between the selenium status and the low T3 syndrome after major trauma", Berger MM, Lemarchand-Beraud T, Cavadini C, Chiolero R., *Intensive Care Med.*, 1996; 22: 575-581.

[48] "Relationship of the increased serum interleukin-6 concentration to changes of thyroidal function in nonthyroidal illness", Bartelena L, Brogioni S, Grasso L, et al., *J. Endocrinol. Invest.* 1994; 17: 269-274.

[49] "The effect of glucocorticoid administration on human pituitary secretion of thyrotropin and prolactin", Re RN, Kourides IA, Ridgway EC, Weintraub BD, Maloof F., *J. Clin. Endocrinol. Metab.*, 1976; 43: 338–347.

[50] "The role of glucocorticoids in the regulation of thyrotropin", Brabant A, Brabant G, Schuermeyer T, et al., *Acta Endocrinol.* (Copenh)., 1989; 121: 95–100.

[51] "Pulsatile thyrotropin secretion in patients with Addison's disease during variable glucocorticoid therapy", Hangaard J, Andersen M, Grodum E, Koldkjaer O, Hagen C., *J. Clin. Endocrinol. Metab.*, 1996; 81: 2502–2507.

[52] "Acute changes in serum thyrotrophin in treated Addison's disease", Ismail AA, Burr WA, Walker PL., *Clin. Endocrinol.* (Oxf)., 1989; 30: 225–230.

[53] "Thyroid hormone concentrations, disease, physical function, and mortality in elderly men", van den Beld AW, Visser TJ, Feelders RA, Grobbee DE and Lamberts SW., *Journal of Clinical Endocrinology and Metabolism*, 2005; 90:6403–6409.

[54] "Cytokines modulate type I iodothyronine deiodinase mRNA levels and enzyme activity in FRTL-5 rat thyroid cells", Pekary AE, Berg L, Santini F, Chopra I, Hershman JM., *Mol. Cell Endocrinol.*, 1994; 101: R31–R35.

[55] "Serum thyroid hormone binding inhibitor in nonthyroid illnesses", Chopra IJ, Huang TS, Beredo A, Solomon DH, Chua Teco GN., *Metabolism*, 1986; 35: 152–159.

[56] "In search of an inhibitor of thyroid hormone binding to serum proteins in nonthyroid illnesses", Chopra IJ, Chua Teco GN, Nguyen AH, Solomon DH., *J. Clin. Endocrinol. Metab.*, 1979; 49: 63–69.

[57] "UDP-glucoronosyltransferases", Burchell B and Coughtrie MW., *Pharmacol. Ther.*, 1983; 43: 261-289.

[58] "Textbook of endocrinology", Larser PR and Ingbar SH, 8th Ed., 1992; 357-487, WB Saunders Co. Philadelphia.

[59] "Pathways of thyroid hormone metabolism", Visser TJ, *Acta Med. Austriaca*, 1996; 23: 10-16.

[60] "Hypothyroidism complicating pregnancy", Davis LE, Leveno KJ, Cunningham FG, *Obstet. Gynecol.*, 1988; 72: 108-12.

[61] "Localization of NADPH-diaphorase reactivity in the chick and mouse thyroid gland", Syed MA, Leong SK, Chan AS, *Thyroid,* 1994; 4: 475-8.

[62] "Role of nitric oxide in the systemic circulation of conscious hyper- and hypothyroid rats", Vargas F, Montes R, Sabio JM, Garcia-Estan J, *Gen. Pharmacol.*, 1994; 25: 887-91.

[63] "Changes in the immunohistochemical localisation of fibroblast growth factor-2, transforming growth factor- beta 1 and thrombospondin-1 are associated with early angiogenic events in the hyperplastic rat thyroid", Patel VA, Hill DJ, Eggo MC, Sheppard MC, Becks GP, Logan A, *J. Endocrinol.*, 1996; 148: 485-99.

[64] "Perinatal outcome in hypothyroid pregnancies", Leung AS., Millar LK., Koonings PP., Montoro M., Mestman JH., *Obstet .Gynecol.*, 1993; 81: 349-53.

[65] "Subclinical hypothyroidism and pregnancy outcomes", Casey BM, Dashe JS, Wells CE, McIntire DD, Byrd W, Leveno KJ, Cunningham FG, *Obstet. Gynecol.*, 2005; 105: 239-45.

[66] "Low birth weight and preeclampsia in pregnancies complicated by hyperthyroidism", Millar LK, Wing DA, Leung AS, Koonings PP, Montoro MN, Mestman JH., *Obstet. Gynecol.*, 1994; 84: 946–949.

[67] "Hypothyrodism and seizures during pregnancy", Mayer DC, Thorp J Baucom D, Spielman FJ, *Am.J.Perinatol.*,1995 May; 12(3): 192-4.

[68] "Interleukin 1β, tumor necrosis factor-α and interleukin 6 decrease nuclear thyroid hormone receptor capacity in a liver cell line", Matthias Wolf, Nils Hansen and Heiner Greten, *European Journal of Endocrinology*, 1994; 131(3), 307-312.

[69] "Werner and Ingbar's the thyroid: a fundamental and clinical text", Sidney C. Werner, Sidney H. Ingbar, Lewis E. Braverman, Robert D. Utiger, Lippincott Williams and Wilkins, 2005, p253.

[70] "Endocrine effects of antiepileptic drugs", Leśkiewicz M, Budziszewska B, Lasoń W., *Przegl Lek*, 2008; 65(11): 795-8 [Article in Polish].

[71] "Antiepileptic Drugs and Thyroid Function", Alberto Verrotti, Alessandra Scardapane, Rossella Manco and Francesco Chiarelli, Journal of Pediatric *Endocrinology and Metabolism*, 2008; 21: 401-8.

[72] "Depression of the serum protein-bound iodine level by diphenyly-dantoin", Oppenheimer JH, Fischer LV, Nielson KM, Jailer JW, *J. Clin. Endocrinol. Metab.*, 1961; 21: 252-262.

[73] "Alteration of thyroid hormone homeostasis by anti-epileptic drugs in humans: involvement of glucuronosyl-transferase induction", Strolin Benedetti M, Whomsley R, Baltes E, Tonner F, *Eur. J. Clin. Pharmacol.*, 2005; 61: 863-872.

[74] "Thyroid function tests in patients on long-term treatment with various anticonvulsant drugs", Liewendahl K, Majuri H, Helenius T, *Clin. Endocrinol. (Oxf)*, 1978; 8: 185-191.

[75] "The effect of thyroxine on magnesium requirement", Vitale J. J., Hegsted D. M., Nakamura M and Connors P, *J. Biol. Chem.*, 1957; 226: 597.

[76] "Magnesium metabolism in hyperthyroidism", Doe R P, Flink E B, Prasad A. S., *J. Lab. Clin. Med.*, 1959; 54: 805.

[77] "Trace elements in hashimoto thyroiditis patients with subclinical hypothyroidism", Erdal M, *Biol. Trace Elem. Res.*, 2008; 123(1-3): 1-7.

[78] "Plasma and erythrocyte magnesium concentrations in thyroid disease: relation to thyroid function and the duration of illness", Yuhei Shibutani et al., *Jpn. J. Med.*, 1989; 28(4): 496-502.

[79] "Magnesium metabolism in hyperthyroidism and hypothyroidism", Jones E. Jones et al., *J. Clin. Invest.*, 1966; 45(6); 891-900.

[80] "Clinical significance of serum magnesium concentration in thyrotoxicosis", Wuttke H, Kessler FJ, *Med. Klin.*, 1976; 71(6): 235-8.

[81] "Effect of aldosteron on the metabolism of magnesium", Horton R, Biglieri E, *J. Clin. Endocrinol. Metab.*, 1972; 22: 1187.

[82] "The ultrafiltrable serum magnesium in hyperthyroidism", Cope CL, Wolff B, *Biochem. J.,* 1942; 36: 413.

[83] "Serum ionized magnesium levels in normal and preeclamptic gestation", Standley CA, Whitty JE, Mason BA, Cotton DB, *Obstet. Gynecol.*, 1997; 89(1): 24-7.

[84] "Serum calcium and serum magnesium in normal and preeclamptic pregnancy", Sukonpan K, Phupong V, *Arch. Gynecol. Obstet.*, 2005; 273(1): 12-6.

[85] "Plasma and intracellular Mg^{2+} concentrations in pre-eclampsia", Kisters K, Niedner W, Fafera I, Zidek W, *J. Hypertens.*, 1990; 8(4): 303-6.

[86] "Blood concentrations of calcium and magnesium in women with severe pre-eclampsia", Villanueva LA, Figueroa A, Villanueva S, *Ginecol. Obstet. Mex.*, 2001; 69: 277-81.

[87] "Intracellular and extracellular, ionized and total magnesium in pre-eclampsia and uncomplicated pregnancy", Sanders R, Konijnenberg A, Hujigen HJ, Wolf H, Boer K, Sanders GT, *Clin. Chem. Lab. Med.*, 1999; 37(1): 55-9.

[88] "The pathogenesis of eclampsia: the 'magnesim-ischaemia' hypothesis", Newman JC, Amarasingham JL, Med. *Hypotheses,* 1993; 40(4): 250-6.

[89] "Magnesium transport in hypertension", Sontia B, Touyz RM, *Pathophysiology*, 2007; 14(3-4): 205-11.

[90] "Intracellular magnesium deficiency and effect of oral magnesium on blood pressure and red cell sodium transport in diuretic-treated hypertensive patients", Hattori K, Saito K, Sano H, Fukuzaki H, *Jpn. Circ. J.*, 1988; 52(11): 1249-56.

[91] "Effects of magnesium on blood pressure and intracellular ion levels of Brazilian hypertensive patients", Sanjuliani AF, de Abreu Fagundes VG, Francischetti EA, *Int. J. Cardiol.*, 1996; 56(2): 177-83.

[92] "Effect of magnesim on red blood cell deformability in pregnancy", Schauf B, Becker S, Abele H, Klever T, Wallwiener D, Aydeniz B, *Hypertens Pregnancy*, 2005; 24(1): 17-27.

[93] "Evaluation of red blood cell deformability and uterine blood flow in pregnant women with preeclampsia or iugr and reduced uterine blood flow following the intravenous application of magnesium", Schauf B, Mannschreck B, Becker S, Dietz K, Wallwiener D, Aydeniz B, *Hypertens Pregnancy*, 2004; 23(3): 331-43.

[94] "Blood rheology in vitro and in vivo", Lowe GD, *Baillieres Clin. Haematol.*, 1987; 1(3): 597-636.

[95] "Magnesium deficiency: pathophysiologic and clinical overview", Al-Ghamdi SM, Cameron EC, Sutton RA, *Am. J. Kidney Dis.*, 1994; 24(5): 737-52.

[96] "Effect of magnesium sulphate on the osmotic fragility and lipid peroxidation of intact red blood cells from pregnant women with severe preeclampsia", Cilia Abad et al., *Hypertension in pregnancy*, 2010; 29(1): 38-53.

[97] "Alterations in osmotic fragility of the red blood cells in hypo- and hyperthyroid patients", Asl SZ, Brojeni NK, Ghasemi A, Faraji F, Hedayati M, Azizi F, *J. Endocrinol. Invest.*, 2009; 32(1): 28-32.

[98] "Principles and practise of endocrinology and metabolism", Kenneth L. Becker, 3rd edit. Lippincot Williams and Wilkins, 2001; p.1932.

[99] "Whole blood viscosity in preeclampsia", Hobbs JB, Oats JN, Palmer AA, Long PA, Mitchell GM, Lou M, Mciver MA, *Am. J. Obstet. Gynecol.*, 1982; 142(3): 288-92.

[100] "Blood viscosity, hyperviscosity and hyperviscosaemia", Leopold Dintenfass, (Kluwer) Springer, 1985, p. 241.

[101] "Is there hyperviscosity in preeclampsia?", Pepple DJ, Reid HL, Mullings AM, *West Indian Med., J* 2000; 49: 229-31.

[102] "Blood viscosity, hyperviscosity and hyperviscosaemia", Leopold Dintenfass, (Kluwer) Springer, 1985, p. 169.

[103] "Hypothyroidism and the influence on human blood rheology", Költringer P, Eber O, Wakonig P, Klima G, Lind P, *J. Endocrinol. Invest.*, 1988; 11(4): 267-72.

[104] "Plasma viscosity in female patients with hypothyroidism: effects of oxidative stress and cholesterol", Konukoglu D, Ercan M ,Hatemi H, Clin. *Hemorheol. Microcirc.*, 2002; 27(2): 107-13.

[105] "Reversal of increased whole blood and plasma viscosity after treatment of hypothyoidism in man", Larsson H, Valdemarsson S, Hedner P, Odeberg H, *Acta Med. Scand.*, 1985; 217(1): 67-72.

[106] "Leukocyte, selenium, zinc and copper concentrations in preeclamptic and normotensive pregnant women", Mahomed K, Williams MA, Woelk Gb et al., *Biol. Trace Elem. Res.*, 2000; 75: 107-118.

[107] "Zinc, cadmium, and hypertension in parturient women", Lazebnik N, Kuhnert BR, Kuhnert PM, *Am. J. Obstet. Gynecol.*, 1989; 161(2): 437-40.

[108] "Elevated copper and lowered zinc in the placentae of pre-eclamptics", Brophy MH, Harris NF, Crawford IL, *Clin. Chim. Acta*, 1985; 145(1): 107-11.

[109] "Plasma and erythrocyte zinc and birth weight in pre-eclamptic pregnancies", Lao TT, Chin RK, Mak YT, Swaminathan R, Lam YM, *Arch. Gynecol. Obstet.*, 1990; 247(4): 167-71.

[110] "Plasma zinc concentration and thyroid function in hyperemetic pregnancies", Lao TT, Chin RK, Mak YT, Panesar NS, *Acta Obstet. Gynecol. Scand.*, 1988; 67(7): 599-604.

[111] "A zinc finger transcription factor ZFAT controls autoimmune thyroid diseases through regulating the expression of immune related genes", Sasazuki Takehiko, *Proceedings of the Japanese Society of Immunology*, 2006; 36: 9.

[112] "Zinc metabolism in thyroid disease", Yoshikazu Nishi, Ryoso Kawate, *Postgraduate Medical Journal*, 1980; 56: 833-837.

[113] "Zinc sulphate supplementation improves thyroid function in hypozincemic Down children", Ines Bucci et al., *Biol. Trace Element Research*, 1998; 67: 3.

[114] "Zinc in endocrinology", Masayuki Kaji, *International Paediatrics*, 2001; 16(3): 1-6.

[115] "The implications of hypozincemia in pregnancy", Adeniyi FA, *Acta Obstet. Gynecol. Scand.*, 1987; 66(7): 579-82.

[116] "Newborn birth weight correlates with placental zinc, umbilical insulin-like growth factor I, and leptin levels in preeclampsia", Diaz E, Halhali A, Luna C, Diaz L, Avila E, Larrea F, *Ach. Med. Res.*, 2002; 33: (1)40-7.

[117] "Zinc supplementation during pregnancy: a double blind randomised controlled trial" Jonsson B, Hauge B, Larsen MF et al. *Acta Obstet. Gynecol. Scand.*, 1996; 75: 725-729.

[118] "The changes of trace elements, malondialdehyde levels and superoxide dismutase activities in pregnancy with or without preeclampsia", Ilhan N, Ilhan N, Simsek M, *Clin. Biochem.*, 2002; 35(5): 393-7.

[119] "Comparison of serum copper, zinc, calcium, and magnesium levels in preeclamptic and healthy pregnant women, Kumru S, Aydin S, Simsek M et al,, *Biol. Trace Elem. Res.*, 2003; 94: 105-12.

[120] "Correlation between maternal plasma homocysteine and zinc levels in preeclamptic women", Harma M, Harma M, Kocyigit A, *Biol. Trace Elem. Res.*, 2005; 104(2): 97-105.

[121] "Serum iron and copper status and oxidative stress in severe and mild preeclampsia", Serdar Z, Gür E, Develioğlu O., *Cell Biochem. Funct.*, 2006; 24(3): 209-15.

[122] "Nonglycosylated ferritin predominates in the circulation of women with preeclampsia but not intrauterine growth restriction", Carl A. Hubel et al., *Clinical Chemistry*, 2004; 50: 948-951.

[123] "Abnormal iron parameters in the pregnancy syndrome preeclampsia", Rayman MP et al., *Am. J. Obstet. Gynecol.*, 2002; 187(2): 412-8.

[124] "Evaluation of the changes in serum iron levels in pre-eclampsia", Shalini Gupta, *Indian Journal of Clinical Biochemistry*, 1997; 12(1): 91-94.

[125] "Third trimester iron status and pregnancy outcome in non-anaemic women ; pregnancy unfavourably affected by maternal iron excess", Lao TT, Tam K-F, Chan LY, *Hum. Reprod.*, 2000; 15: 1843-1848.

[126] "A conservative approach to iron supplementation during pregnancy", Graves BW, Barger MK, *Midwifery Womens Health*, 2001; 46(3): 159-66.

[127] "Role of iodine, selenium and other micronutrients in thyroid function and disorders", Triggiani V et al., Endocr Metab Immune Disord. *Drug Targets*, 2009; 9(3): 277-94.

[128] "Low selenium status is associated with the occurrence of the pregnancy disease preeclampsia in women from the United Kingdom", Rayman MP, Bode P,Redman CW, *Am. J. Obstet. Gynecol.*, 2003; 189(5): 1343-9.

[129] "Magnesium supplementation during pregnancy: a double-blind randomized controlled clinical trial", Sibai BM, Villar MA, Bray E, *Am. J. Obstet. Gynecol.*, 1989; 161(1): 115-9.

[130] "Thyroid autoimmunity and its association with non-organ-specific antibodies and subclinical alterations of thyroid function in women with a history of pregnancy loss or preeclampsia", Mecacci F, Parretti E, Cioni R, Lucchetti R, Magrini A, La Torre P, Mignosa M, Acanfora L, Mello G, *J. Reprod. Immunol.*, 2000 Feb; 46(1): 39-50.

[131] "Thyroid stimulating hormones levels in cord blood are not influenced by non-thyroidal mother' s diseases", Ward LS, Kunii IS, de Barros Maciel RM, *Sao Paulo Med. J.*, 2000 Sep 7; 118(5): 144-7.

[132] "Cord blood thyroid-stimulating hormone levels in high-risk pregnancies", Chan LY, Chiu PY, Lau TK, *Eur. J. Obstet. Gynecol. Reprod. Biol.*, 2003 Jun 10; 108(2):142-5.

[133] "Thyroid function tests in preterm infants born to preeclamptic mothers with placental insufficiency", Belet N, Imdat H, Yanik F, Kücüködük S, *J. Pediatr. Endocrinol. Metabol.*, 2003; 16(8): 1131-5.

[134] "Of pregnancy and progeny", Freinkel N, Banting lecture, *Diabetes*, 1980; 11023-35.

[135] "Thyroid hormone concentrations in preterm infants born to pre-eclamptic women with placental unsufficiency", Fetter WP, Waals-Van de Wai CM, Van Eyck J, Samson G, Bongers--Schokking JJ, *Acta Paediatrica*, 1998; 87(2): 186-190.

[136] "Thyroid function in very low birth weight infants", Klein RZ. Carlton EL., Faix JD, Frank JE, Hermos RJ, Mullaney D et al., *Clin. Endocrinol.* (Oxford), 1997; 47: 411-7.

In: Genetic Diagnoses
Editor: R. J. Sarma, pp. 29-52

ISBN: 978-1-61324-866-9
© 2012 Nova Science Publishers, Inc.

Chapter II

Molecular Diagnosis of Thalassemia

*Wen Wang[1], Boran Jiang[2] and Samuel S. Chong[2,3]**
[1]Department of Obstetrics and Gynecology,
University of California, Irvine, USA
[2]University Children's Medical Institute,
National University Hospital, Singapore
[3]Department of Pediatrics, Yong Loo Lin School of Medicine,
National University of Singapore, Singapore

Abstract

The thalassemias are a group of inherited hematologic disorders characterized by hemolytic anemia. They are among the most common single gene disorders in the world and are especially frequent in South-East Asia, the Middle East, India, the Mediterranean and Africa. Due to demographic changes in the last few decades, the incidence of thalassemia in North European countries and North America has increased. In normal adults, the major hemoglobin (Hb) consists of two α- and two β-globin chains which are covalently linked to heme, the oxygen-binding group. The two major forms of thalassemia, α- and β-thalassemia, are due to absent or reduced production of α-globin and β-globin chains, respectively. While the majority of α-thalassemia is due to single or double gene deletions in the α-globin gene, β-thalassemia is mostly caused by point mutations in the β-globin gene. Affected individuals with different mutations suffer from variable degrees of anemia, which can be detected by hematological/biochemical investigations such as complete blood count (CBC), Hb analysis by isoelectric focusing (IEF) or high-performance liquid chromatography (HPLC). Molecular analysis is usually required to confirm the hematological observations and make a definitive diagnosis of the causative mutations to inform management and counseling. Many molecular diagnostic and screening assays have been developed to identify the mutations in the α- and β-globin genes. This article aims to review the various technical platforms, including PCR

* Samuel S. Chong, Department of Pediatrics, National University of Singapore, Level 12, NUHS Tower Block, 1E Kent Ridge Road, Singapore 119228, SINGAPORE. Phone 65-6772-4152; Fax 65-6772-4100; Email: paecs@nus.edu.sg.

based methods and high-throughput approaches, and common strategies available for the molecular diagnosis of thalassemia. Also discussed are thalassemia prevention strategies such as prenatal and pre-implantation genetic diagnosis.

1. Background

1.1. Epidemiology

The thalassemias (α and β) rank among the most common of single gene disorders globally. They are especially prevalent in South-East Asia, India, the Middle East, the Mediterranean countries, and Africa [1-3]. α-thalassemia is most prevalent within Southeast Asia, while β-thalassemia is common over a wider geographical area including Southeast Asia, India, the Middle East and the Mediterranean.

Due to recent migration patterns over the last few decades, today's epidemiology of thalassemia is less geographically restrained than in the past. For example, the incidence of thalassemia in northern European countries and in North America has increased due to large inflows of immigrants from thalassemia prevalent regions. These demographic changes have resulted in thalassemia becoming a significant health problem in much of the Western hemisphere [3, 4].

Interestingly, thalassemia occurs at high frequencies throughout all tropical and subtropical regions of the world where malaria is also endemic, suggesting a functional association between the presence of globin gene mutations and a selective survival advantage against malaria [5]. The mechanisms underlying this protection remain incompletely understood.

1.2. Pathophysiology

The thalassemias are a heterogeneous group of genetic disorders characterized by a reduced or absent production of normal hemoglobin (Hb), thus resulting in hypochromic anemia [6]. The two most clinically relevant thalassemia syndromes are α- and β-thalassemia. α-thalassemia is caused by deletions or point mutations affecting either one or more of the duplicated α-globin genes located on chromosome 16p13.3, leading to absent or decreased α-globin chain production. In contrast, β-thalassemia results from point mutations or more rarely deletions of the β-globin gene located on chromosome 11p15.5, leading to absent or decreased β-globin chain production.

1.2.1. Hemoglobin Component and Types

The constitution of hemoglobin (Hb) is different at different stages of human development, although all normal hemoglobins have a similar tetrameric structure that consists of one pair each of α-like and β-like globin chains. At up to eight weeks of gestation, three types of embryonic hemoglobins are present, Hb Gower 1 ($\zeta 2\varepsilon 2$), Hb Gower 2 ($\alpha 2\varepsilon 2$), and Hb Portland ($\zeta 2\gamma 2$). At about five weeks of gestation, the embryonic hemoglobins start to be replaced by fetal hemoglobin, Hb F ($\alpha 2\gamma 2$). Hb F is the predominant hemoglobin in the developing fetus until birth, when it is replaced by the adult hemoglobins HbA ($\alpha2\beta2$) and HbA2 ($\alpha2\delta2$). Approximately 95% of all Hb in a healthy adult is HbA, while HbA2 constitutes less than 3.2%, and less than 1% is HbF.

1.2.2. α-Thalassemia

1.2.2.1. α-Globin Gene Cluster

The α-globin gene cluster is located on chromosome 16p13.3 and includes seven genes and pseudogenes in the following order: 5'-ζ-$\psi\zeta$-μ-$\psi\alpha1$-$\alpha2$-$\alpha1$-$\theta1$-3'. These genes are under the regulatory control of the α-globin HS-40 enhancer [7, 8] (Figure 1). The cluster contains three genes of functional significance (ζ, $\alpha2$ and $\alpha1$) and two pseudogenes ($\psi\zeta$ and $\psi\alpha1$). A third pseudogene $\psi\alpha2$ has now been shown to be expressed and is now referred to as μ or α^D [9, 10], while $\theta1$ is a functional gene of unknown significance. The ζ-globin gene (OMIM 142310) is the embryonic α-like gene, expressing up to eight weeks of gestation. There is no fetal α-like gene, and thus the $\alpha2$ (OMIM 141850) and $\alpha1$ (OMIM 141800) genes begin to be expressed from the eighth week of gestation. They are highly homologous, differing in sequence only within the second intron and 3' untranslated region.

They thus encode identical proteins, although the expression level of these two genes differs, as $\alpha2$ produces 2-3 times more α-globin chain than $\alpha1$ [11]. This different expression of $\alpha2$ and $\alpha1$ genes has important clinical implications, as the effect of a mutation on α-globin chain production will depend on which α-globin gene the mutation resides in. Reduced or absent α-globin chain production due to $\alpha2$ and/or $\alpha1$ gene mutations causes an imbalance in the ratio of α-like globin to β-like globin chains, leading to the formation of non-functional and unstable homo-tetramers of β-like chains such as Hb H ($\beta4$) and Hb Bart's ($\gamma4$).

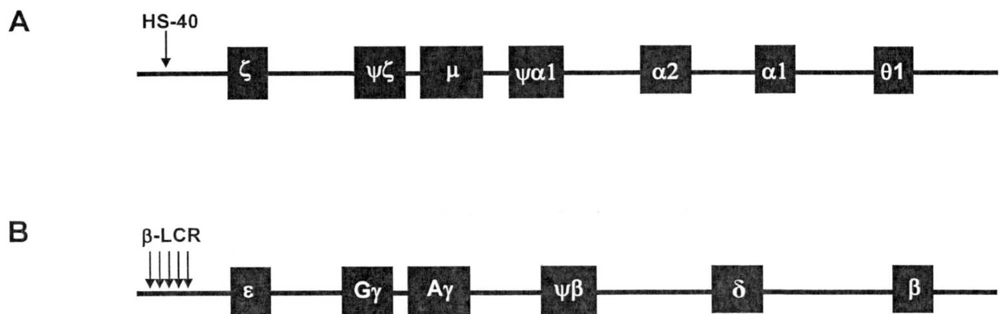

Figure 1. The α- and β-like globin gene clusters. (A) The α locus on chromosome 16p13.3. Of the seven genes shown only three genes are expressed at a clinically significant level, ζ (expressed in early fetal development) and $\alpha1$ and $\alpha2$. The other genes shown are pseudogenes ($\psi\zeta$ and $\psi\alpha1$), a previously known pseudogene $\psi\alpha2$ (now has been shown to be expressed and referred as μ or α^D), or is not expressed at a significant level ($\theta1$). The vertical arrow indicates the location of the upstream hypersensitive site (HS-40), the regulator for locus gene expression. (B) The β locus on chromosome 11p15.5, with the ε, Gγ and Aγ, δ and β genes, arranged in the order of their developmental expression. ε globin switches to the Gγ and Aγ globin genes at 6–8 weeks gestation, and the second switch from the γ genes to the δ and β genes occurs shortly prior to birth. $\psi\beta$ is a non-expressed pseudogene. Upstream of the β-globin cluster is the β locus control region (β-LCR) consisted of the five hypersensitive sites indicated by vertical arrows.

1.2.2.2. Classifications of α-Thalassemia

Based on the number of functional α-globin gene copies inactivated or deleted, individuals can be categorized into several α-thalassemia groups:

1) Silent carriers (heterozygous α-thalassemia-2): where a single α-globin gene is mutated/deleted on one chromosome 16;
2) α-thalassemia trait: characterized by mutation/deletion of two α-globin genes, either *in cis* (heterozygous α-thalassemia-1) or *in trans* (homozygous α-thalassemia-2);
3) Hb H disease: characterized by mutation/deletion of two α-globin genes *in cis* on one chromosome 16 and a third α-globin gene on the other chromosome 16;
4) Hb Bart's hydrops fetalis syndrome: characterized by mutation/deletion of all four α-globin genes.

The symptoms of the above vthalassemia subtypes range from nil (silent carriers) to mild anemia (α-thalassemia trait); transfusion independent or dependent severe anemia (Hb H disease), and *in utero* or neonatal demise (Hb Bart's hydrops fetalis).

1.2.2.3. Molecular Etiology of α-Thalassemia

Most of the common α-thalassemia mutations involve deletions, gross rearrangements, or unequal crossover of large genomic fragments containing the genes and their regulatory sequences [6]. The most common α-thalassemia deletions include the Southeast Asian ($--^{SEA}$, $--^{FIL}$, $--^{THAI}$) and Mediterranean ($--^{MED}$, $-(\alpha)^{20.5}$) double-gene deletions *in cis*, and the 3.7 kb rightward ($-\alpha^{3.7}$) and 4.2 kb leftward ($-\alpha^{4.2}$) single-gene deletions (Figure 2).

Less frequent are the point mutations which can be found in critical regions of the α-globin gene. More than 45 different non-deletional determinants of α-thalassemia have been described, and the HbVar web-site (http://globin.bx.psu.edu/hbvar/) maintains an updated list of all known mutations [12, 13]. Many mutations have been described affecting mRNA processing, mRNA translation, or α-globin stability. The most common of these mutations include Hb Constant Spring (HbCS) and Hb Pakse, whose termination codon mutations lead to elongated Hb variants, and Hb Quong Sze and Hb Suan Dok, whose mutations lead to highly unstable structural variants. Hb H disease can be caused by compound heterozygosity for a double-gene deletion on one chromosome and a point mutation of one α-globin gene on the other chromosome ($--/\alpha^{T}\alpha$) [14], and individuals with this genotype tend to experience a more severe clinical course compared with Hb H disease individuals who are compound heterozygous for a double-gene deletion and a single gene deletion ($--/-\alpha$). It is thought that whereas deletion of one α-globin gene leads to an increase in expression of the remaining α-globin gene *in cis*, such a compensatory upregulation of the normal gene copy does not occur when its adjacent gene copy is inactivated by a point mutation.

1.2.3. β-Thalassemia

More than 200 different β-thalassemia causative mutations have been described [15], the overwhelming majority of which are single-nucleotide substitutions or insertions/deletions of single nucleotides or oligonucleotides that lead to a shift in the peptide reading frame.

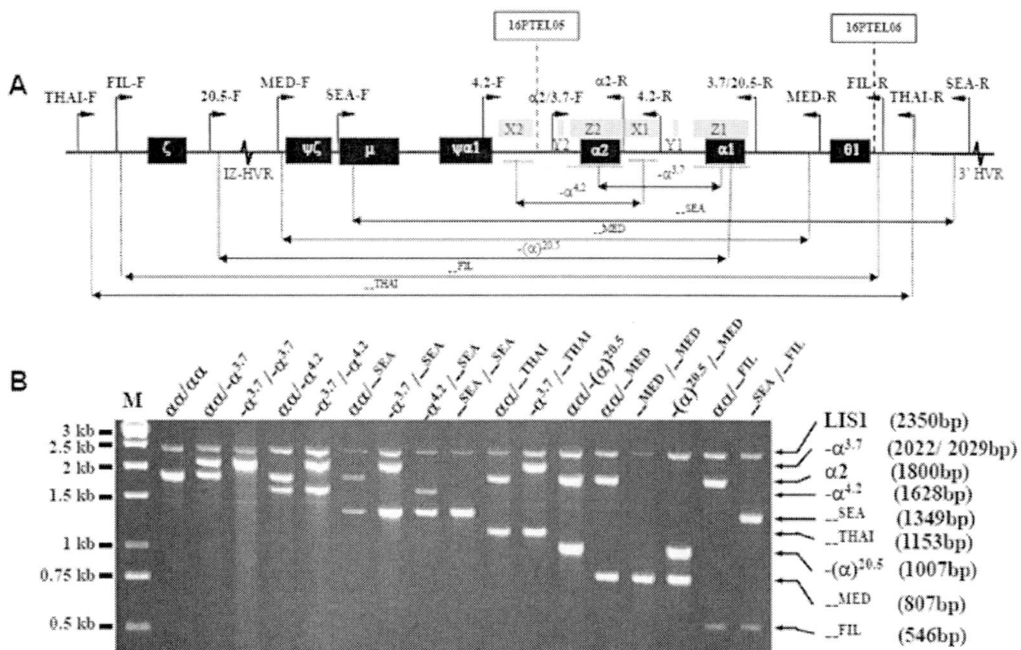

Figure 2. Screening of common deletional determinants of α-thalassemia by multiplex polymerase chain reaction. (A) The α-globin gene cluster, indicating extents of 7 common deletions and relative positions of the primers. (B) Multiplex PCR analysis of DNA samples with previously determined α-globin genotypes. For further details of the multi-mutation screening assay, please refer to [28].

1.2.3.1. β-Globin Gene Cluster

The β-globin gene cluster is located on chromosome 11p15.5 and includes five functional genes and one pseudogene in the following order: 5'-ε-Gγ-Aγ-ψβ-δ-β-3'. Several short tandem CA repeats and ATTTT repeats occur between the δ- and β-globin genes, and numerous single nucleotide polymorphisms (SNPs) occur within the β-globin gene cluster [6]. The functional genes are under the control of the β-globin locus control region (β-LCR) [16] (Figure 1) and arranged in the order of their developmental expression. ε globin (OMIM 142100) switches to the Gγ (OMIM 142250) and Aγ (OMIM 142200) at 6–8 weeks gestation, and the second switch from the γ-globin genes to the δ- (OMIM 142000) and β- (OMIM 141900) globin genes occurs shortly prior to birth. ψβ is a non-expressed pseudogene.

1.2.3.2. Classifications of β-Thalassemia

Based on the genotypes and clinical symptoms, β-thalassemia can be categorized into three groups:

1) β-thalassemia minor (trait): these individuals are carriers of a mild β^+ mutation (heterozygous β^+ thalassemia) or of a more severe β^0 mutation (heterozygous β^0 thalassemia). β^+ mutations usually occur within the promoter, at a splice-site consensus sequence, or within the polyadenylation signal or untranslated regions, and lead to reduced production of the β-globin peptide from the mutated gene copy. β^0 mutations usually lead to non-functional mRNAs, and usually take the form of nonsense or frameshift mutations at the initiation codon or elsewhere within the

coding region, although splice-site consensus sequence mutations can also lead to complete loss of β-globin peptide production from the mutated gene copy;

2) β-thalassemia intermedia: individuals are either homozygous for two β^+ mutations thalassemia or compound heterozygous for a β^+ mutation and a β^0 mutation;

3) β-thalassemia major (Cooley's anemia): individuals are either compound heterozygous for a β^+ mutation and a β^0 mutation or are homozygous for two β^0 mutations.

With only one affected β-globin gene, individuals with β-thalassemia minor (trait) are usually clinically asymptomatic, although hematologic analysis may reveal microcytosis and mild anemia. Individuals with β-thalassemia major (Cooley's anemia) produce very little or no β-globin peptide, and suffer from severe anemia and hepatosplenomegaly. They usually come to medical attention within the first two years of life. Without blood transfusions and related medical treatment, affected children suffer from severe failure to thrive and a very short life expectancy. Individuals with β-thalassemia intermedia produce significantly reduced β-globin chains, with symptoms appearing from 2-6 years of life to adulthood. Symptoms vary in severity from mild anemia to moderate anemia not requiring lifelong blood transfusions, and affected individuals can survive well beyond 20 years of age.

Genotype-phenotype correlations of β-thalassemia mutations are not as clear as that for α-thalassemia. For example, compound heterozygosity for a β^+ and a β^0 mutation produces a variable phenotype ranging from thalassemia intermedia to thalassemia major. This is partly due to the observation that clinical severity of β-thalassemia depends on the extent of α-globin chain/non-α-globin chain imbalance. Concomitant reduction in the α-globin chain production reduces this α -globin chain/non-α-globin chain imbalance, resulting in a mild β-thalassemia phenotype. Additionally, while β-thalassemia has long been regarded as a recessive disorder, heterozygous carriers of certain nonsense, frameshift, missense, or insertion/deletion mutations have been reported to be affected by dominant β-thalassemia, characterized by a thalassemia intermedia phenotype of variable clinical severity.

1.2.3.3. Molecular Etiology of β-Thalassemia

β-thalassemia mutations are very heterogeneous, with over 200 disease-causing β-globin gene mutations having been identified to date. Despite this heterogeneity, the common mutations are generally limited within each affected ethnic population. For example, in Southeast Asia and India, three mutations, IVSI,5 G→C, codon41/42 −TTCT and IVSI,1 G→T, account for more than two thirds of the β-thalassemia mutant alleles in these two regions.

2. Laboratory Testing of Thalassemia

Laboratory tests play an important role in the diagnosis of thalassemia. Improvements in diagnostic technologies have improved the utility and reliability of these tests in confirming a clinical diagnosis, explaining the hematological abnormality, prenatal diagnosis, genetic counseling and carrier screening. Laboratory tests can be subdivided into hematologic/biochemical and molecular. Hematologic/biochemical tests include a complete blood count (CBC), hemoglobin (Hb) H inclusion body staining, high performance liquid chromatography (HPLC) detection and quantification of HbA2 and HbF, and Hb electrophoresis and isoelectric focusing. Molecular methods rely on DNA analysis to identify the underlying mutations. A definitive diagnosis of the disease-causing mutations is important for disease management, genetic counseling and reproductive decision-making. The following sections describe currently available methods for the laboratory diagnosis of thalassemia, with an emphasis on molecular diagnostic techniques.

2.1. Hematologic and Biochemical Screening

2.1.1. Complete Blood Count (CBC)

CBC analysis provides information about red blood cell (RBC) count, mean corpuscular volume (MCV), mean corpuscular hemoglobin (MCH), mean corpuscular hemoglobin concentration (MCHC), red cell distribution width (RDW), hematocrit (HCT), white blood cell (WBC) count, and platelet count.

Generally, thalassemias are characterized by hypochromatic and microcytic anemia. RBC count is increased and the Hb concentration is decreased to varying degrees, concomitant with a low MCV and MCH. Asymptomatic individuals with MCH below 27pg or MCV below 80fl should be investigated for thalassemia trait [6, 17]. HbA2 level is increased in β-thalassemia minor, but variable in β-thalassemia homozygotes [18].

2.1.2. Electrophoretic Analysis

Electrophoresis is a widely used technique for the identification of variant hemoglobins. It can be carried out on filter paper, cellulose acetate membrane, or agarose to separate variant hemoglobins at alkaline pH or acid pH. For example, electrophoresis at acid pH will distinguish HbC from HbE, and HbO/HbS from HbD/HbG. Limitations of this test include the inability to accurately quantify low concentration Hb variants (HbA_2) or detect fast Hb variants (HbH, HbBart's) [17, 19, 20].

Isoelectric focusing (IEF) has been used to separate various hemoglobins in a pH gradient gel according to their isoelectric point (pI). The sharper bands on IEF allow the separation of some hemoglobins that cannot be distinguished by other electrophoretic methods. However, it is also unsuitable for precise quantification of low concentration Hbs (HbA_2) [17, 19]. IEF also can be used in distinguishing between deletional and non-deletional HbH disease, due to the good separation between HbH and HbCS (hemoglobin Constant Spring) tetramers.

2.1.3. High Performance Liquid Chromatography (HPLC)

HPLC is well suited for identification and quantification of normal and variant hemoglobins such as HbA, HbA_2, HbF, HbS, and HbC, and is routinely used to diagnose β-thalassemia and hemoglobinopathies [17, 19].

2.1.4. Staining of HbH inclusion bodies

In HbH disease and some α^0-thalassemia carriers, excess β-globin chains that are unable to form tetramers with α-globin chains will instead form β_4 homotetramers (HbH). HbH is unstable in solution, precipitates readily in red blood cells, and can be visualized under the microscope [17].

2.2. Molecular Methods

Numerous DNA analysis techniques have been developed for the detection of deletional and non-deletional mutations in the α- and β-globin genes. Since deletional mutations account for most of the cases in α-thalassemia and non-deletional mutations lead to most of the cases in β-thalassemia, different diagnostic methods have been developed for the detection of the different deletions and non-deletional mutations present in α- and β-thalassemia. The prevalence of a limited number of mutations within each ethnic group has allowed the development of ethnic-specific diagnostic screening panels targeted to specific populations. For the rarer mutations, and in cases where targeted mutation screens are negative, gene re-sequencing remains the method of choice.

2.2.1. Southern Blot Analysis

This was the main technique for the detection of common α-globin gene deletions, as well as gene triplications and quadruplication until PCR-based methods were developed. Nitrocellulose or nylon blots containing *Bam*HI or *Bgl*II digested patient DNA are hybridized with an α-globin gene probe or a ζ-globin gene probe. This test can detect the $-\alpha^{3.7}$ and $-\alpha^{4.2}$ single-gene deletions, as well as some double-gene deletions such as $-(\alpha)^{20.5}$, $--^{SEA}$ and $--^{MED}$. The $--^{THAI}$ and $--^{FIL}$ deletions, however, are detectable only on blots containing *Sst*I digested DNA that are hybridized with a DNA probe located upstream of the ζ-globin gene (LO probe) [6, 21].

2.2.2. Gap PCR for Deletions

This is a technique that detects the presence of large deletions. Using amplification primers that anneal to regions flanking the deletion, a deletion junction fragment of an expected size is successfully amplified only when the deletion is present, due to the proximity of the opposing primers. Conversely, no amplification product is generated from a non-deleted allele since the opposing primers are too far apart for successful amplification under normal thermal cycling conditions.

Gap PCR assays have been developed to detect the most common single-gene and double-gene deletions [22-27]. A multiplex Gap-PCR assay has been developed to screen for seven common α-thalassemia deletions in one reaction including $--^{SEA}$, $--^{THAI}$, $--^{FIL}$, $-^{MED}$, $-(\alpha)^{20.5}$, $-\alpha^{3.7}$ and $-\alpha^{4.2}$ [28]. The expected amplicon sizes for each of the deletion junction fragments and the positive control $\alpha2$ fragments are different, allowing easy visualization and discrimination of the various junction fragments on an agarose gel (Figure 2). Evaluation of the multiplex-PCR in a clinical laboratory suggests that it is suitable as a standard clinical screening protocol for detecting common α-globin deletions [29]. Another multiplex Gap-PCR assay includes the $--^{SA}$ double-gene deletion that was originally found in a patient from South Africa [30], but is now known to be prevalent in the Indian population [31]. Gap PCR assays for detection of β-thalassemia deletions have also been developed [6, 32].

2.2.3. Dot-Blot and Reverse Dot-Blot Hybridization

Dot-blot and RDB methods have been widely used in the diagnosis of β-thalassemia [33-37] and α-thalassemia [38] point mutations. Both methods are based on the hybridization of allele-specific oligonucleotide (ASO) probes to amplified genomic DNA. In dot-blot hybridization, amplified genomic DNA is spotted onto a nylon membrane and allowed to hybridize with a radio-labeled oligonucleotide probe. Two identical spotted membranes are typically prepared, one for hybridization with a probe complementary to the mutation and the other for hybridization with a probe complementary to the wild-type or normal sequence. When at least one allele in the DNA sample is complementary to the probe, a stable DNA-probe hybrid is created, which yields a black spot upon exposure to autoradiography film. In contrast, when both alleles in the DNA sample show a mismatch with the probe, the probe cannot hybridize stably to the DNA and no signal is produced on the autoradiography film.

In reverse dot-blot hybridization (RDB), multiple ASO probes are affixed onto the membrane, enabling a large number of mutations to be screened in a single hybridization step. Two oligonucleotide probes are synthesized for each mutation site, one that is complementary to the mutant sequence, and the other complementary to the normal sequence. The genotype of the DNA sample is determined by the presence or absence of hybridization signal from the mutant and wild type probe spots.

2.2.4. PCR - Restriction Fragment Length Polymorphism (RFLP) Analysis

This method has been applied to the identification of Mediterranean β-thalassemia mutations [39]. It combines PCR amplification and restriction endonuclease digestion of the amplicon. More than 40 β-thalassemia mutations create or destroy a restriction endonuclease site, but this still represents a small fraction of all known mutations [6]. Incomplete or partial digestion of PCR products can lead to false negative or false positive results; thus positive and negative controls should always be included.

2.2.5. Denaturing Gradient Gel Electrophoresis (DGGE)

DGGE is suitable for screening for the presence of unknown mutations, after which the exact position of the mutation within the fragment can be determined by DNA sequencing. It has been used to screen for β-thalassemia [40-43] and α-thalassemia [44] point mutations. PCR amplification from samples heterozygous for a mutation generates heteroduplex and homoduplex DNA fragments during denaturation and reannealing of complementary single-stranded DNA molecules. Heteroduplex fragments are formed when the reannealed strands contain a single mismatched nucleotide, which affects the stability of the duplex. Heteroduplex fragments migrate differently from homoduplex fragments during electrophoresis through a denaturing gradient gel due to their tendency to strand-dissociate earlier and under milder denaturing conditions compared to homoduplexes. Thus heterozygous samples will generate migration patterns that are different from those of homozygous samples [42].

2.2.6. Amplification Refractory Mutation System (ARMS)

Several ARMS assays have been developed for detection of thalassemia point mutations [45-49]. ARMS is a simple method for targeted identification of mutations. In ARMS, allele-specific primers are designed with their 3' terminal nucleotide complementary to either the mutant or normal nucleotide. Thus when the mutation is present, a PCR reaction containing the mutant-specific primer will successfully amplify a fragment. Conversely, when the mutation is absent in the sample, no PCR product will be generated due to the base-pair mismatch 3' nucleotide of the primer with the DNA template. To enhance the specificity of assay, a deliberate additional mismatch is introduced into the primer sequence 3-4 nucleotides away from the 3'end of the primer. To control for PCR failure, a pair of control primers is included to amplify an unrelated fragment simultaneously [50].

2.2.7. Direct DNA Sequencing

Direct DNA sequencing based on the Sanger sequencing technology has been the gold standard method for detection of known and unknown mutations. This technique has been applied to mutation screening at the α-globin [51, 52] and β-globin [53] gene loci. It has also been used for characterization of mutations detected by DGGE and ARMS [40, 42, 44, 53]. Newer next-generation sequencing technologies, which can generate more sequence data more cheaply and rapidly [54], have also been used in the screening and identification of known and unknown mutations in both α- and β-thalassemias [55, 56].

2.2.8. Minisequencing

In minisequencing, a primer is hybridized to the amplified DNA fragment such that it ends one nucleotide upstream of the position of a mutation site of interest. This is followed by DNA polymerase mediated single base primer extension by a terminator nucleotide complementary to the target site nucleotide. As each of the four terminator nucleotides is labeled with a different fluorophore, the color of the fluorescent tag(s) on the extended primer will indicate the presence of a normal and/or mutant allele in the template DNA.

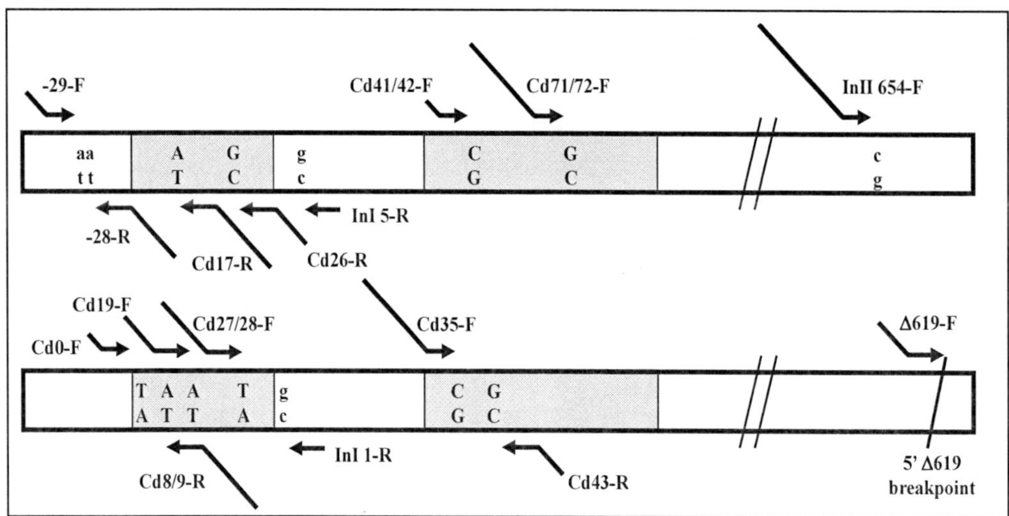

Figure 3. Common mutations in the β-globin gene (HBB) and primers used in the minisequencing screen of these mutations. Each minisequencing primer differs in length from the others by the inclusion of nonspecific tails of varying length at their 5' ends.

Minisequencing can be multiplexed to enable the screening of multiple mutation sites in a single reaction. In multiplex minisequencing, the primers are designed with tails of different lengths, so that they can be separated by size upon capillary electrophoresis [57]. The color and number of the peaks at each mutation site in the electropherogram are then used to determine the patient's genotype. Multiplex-minisequencing has been used in the detection of point mutations in both α- and β-thalassemia.

The β-thalassemia multiplex-minisequencing assay can detect 15 common Southeast Asian and Indian mutations [58]. The assay was divided into two multiplex panels due to the close proximity of the mutations (Figures 3 and 4). The α-thalassemia multiplex-minisequencing assay detects seven point mutations [59] (Figures 5 and 6).

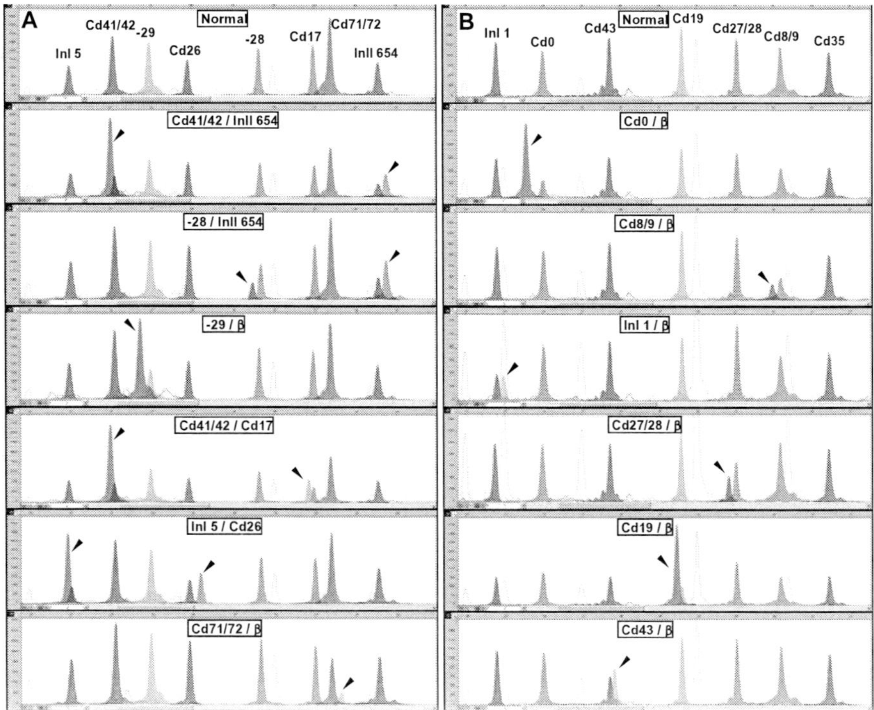

Figure 4. Separation of multiplex minisequencing products by capillary electrophoresis to identify mutations screened for in panels A and B. Shown are electropherograms from a normal individual and patient samples with known thalassemia genotypes. The position of the extended primer peak specifies the mutation locus, whereas the peak color specifies the allele/nucleotide. Each mutant allele displays a characteristic peak color/shade, position, and height relative to its wild-type allele peak. Further details can be found in [58].

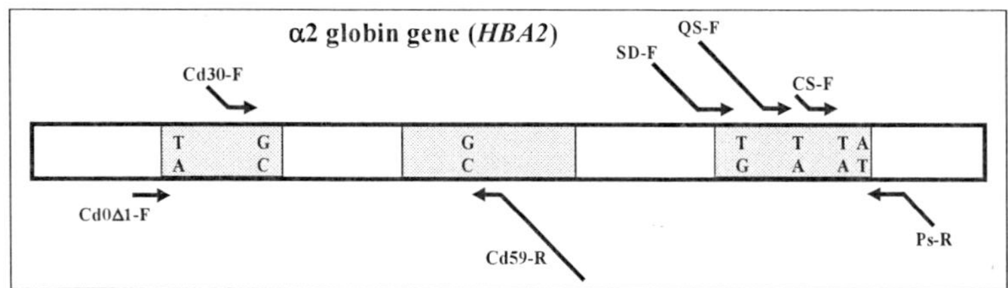

Figure 5. Seven of the more common point mutations within the α globin gene (HBA) and tailed primers used in the minisequencing screen of these mutations.

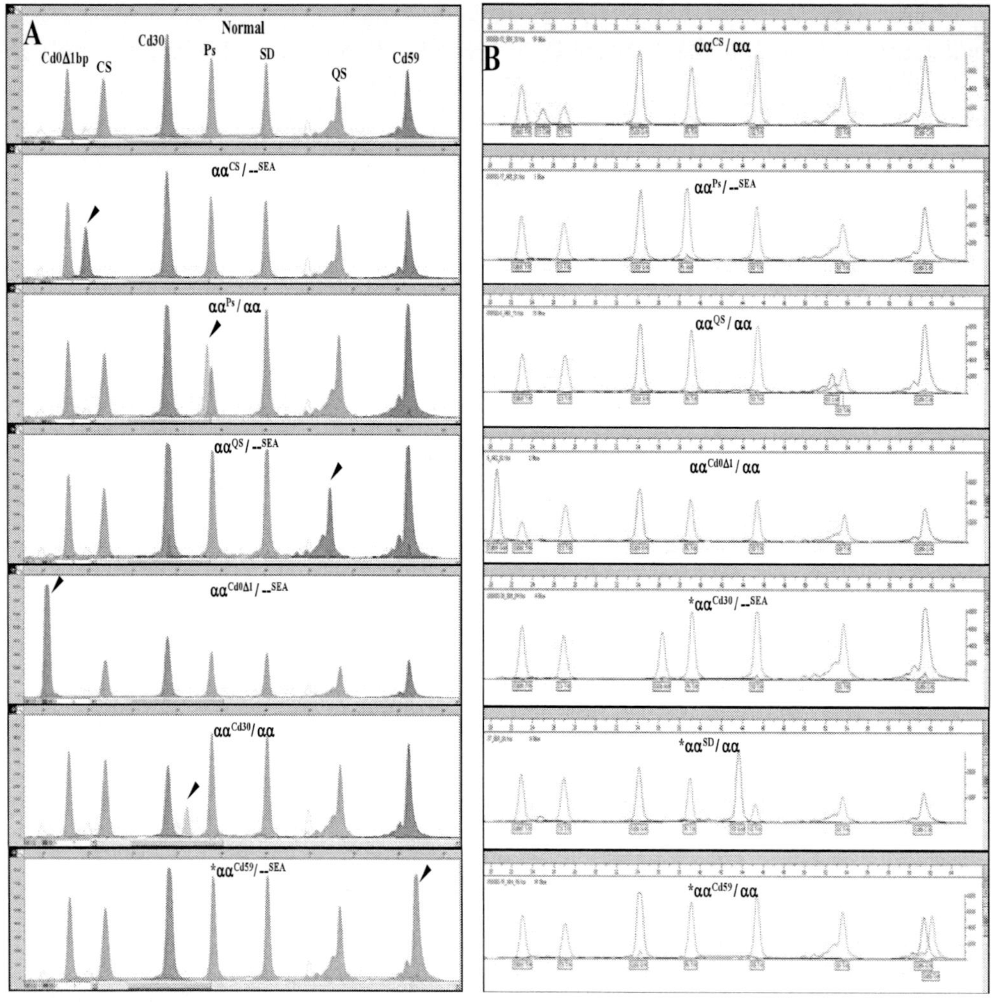

Figure 6. Multiplex minisequencing of α2 globin gene point mutations, visualized with GeneScan™ (A) or Genotyper™ (B). The Genotyper application allows user customization to automatically assign labels under each peak, which can list information on mutation site, nucleotide added (dA, dG, dC, or dT), and wild-type (N) or mutant (M) result status. * indicate reconstituted DNA samples containing a mixture of genomic DNA and a plasmid clone containing the point mutation. Further details can be found in [59].

2.2.9. Multiplex Ligation-Dependent Probe Amplification (MLPA)

MLPA is a multiplex ligation-mediated PCR method generally applicable to the determination of copy number changes at virtually any gene locus. In this assay, paired MLPA probes are designed to be complementary to adjacent segments of a target site. The left half of each probe pair contains a common tail sequence at its 5' end, while the right half of the probe pair contains a common tail at its 3' end with a sequence different from the leftward probe's tail. In the presence of the target sequence, both probes of the probe pair will anneal adjacent to each other, whereupon the probe pair are successfully ligated together and amplified by a pair of PCR primers complementary to their 5' and 3' tails. Probe pairs for up to 50 gene loci can be annealed, ligated, and amplified together in a single multiplex reaction, using a common PCR primer pair. To distinguish the different genetic loci, stuffer sequences of differing length are inserted into the tail of the rightward probe, so that the amplicons can be easily separated by capillary electrophoresis based on size [60]. The MLPA reaction can be divided in five major steps: 1) DNA denaturation and hybridization of MLPA probes; 2) ligation reaction to ligate the hybridized paired probes; 3) PCR reaction to amplify the ligated probes; 4) separation of amplification products by electrophoresis; and 5) data analysis (Figure 7).

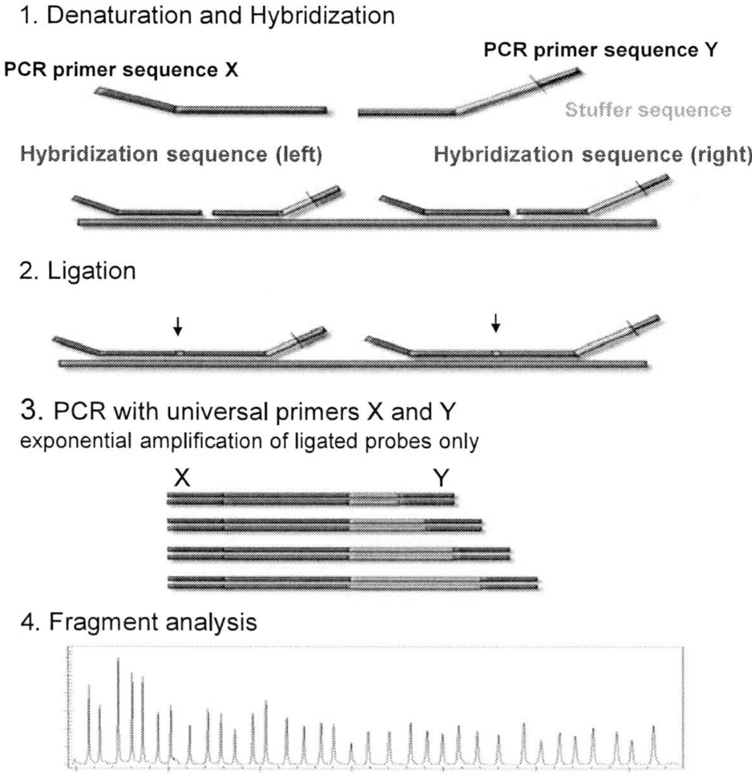

Figure 7. Outline of the MLPA strategy (reproduced from MRC-Holland, www.mlpa.com).

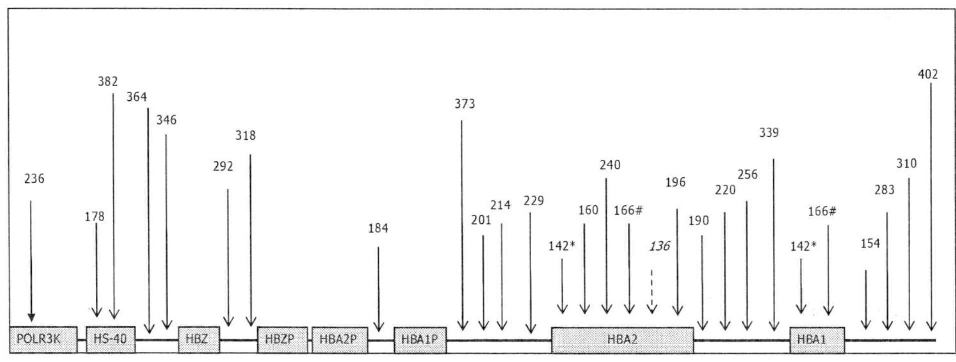

Figure 8. Locations of the probes in the P140-B3 HBA MLPA kit. The numbers above the arrows represent the amplification size of the respective probe in base pairs. *142 nt; # 166 nt: The 142 nt and 166 nt probes both detect a sequence that is present in both HBA1 and HBA2.

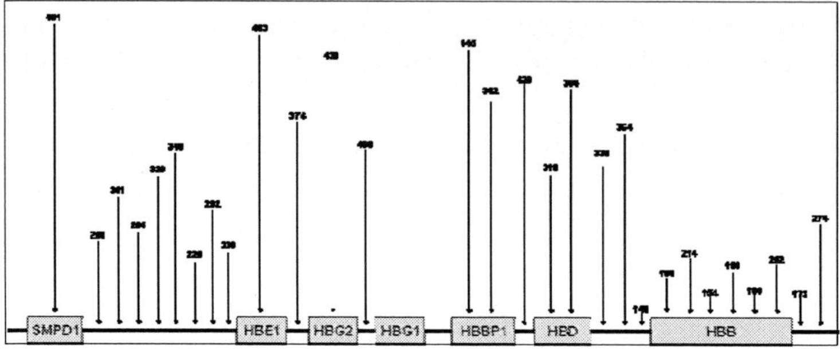

Figure 9. Locations of the probes in the P102-B1 HBB MLPA kit.

MLPA has been applied in the detection of deletions/duplications and non-deletional mutations in both α- and β-globin gene loci [61-64]. In the α-thalassemia MLPA (SALSA MLPA kit P140 HBA, MRC-Holland), there are 24 pairs of probes to cover the HBA gene cluster, including the HS-40 regulatory element and one probe to detect the Constant Spring point mutation (Figure 8). There are 28 probes in the β-thalassemia MLPA (SALSA MLPA kit P102 HBB, MRC-Holland), to cover the HBB gene cluster, including several probes for the upstream regulatory sequences and one probe for the Hb S mutation (Figure 9). This method is rapid, effective, and most importantly, it can detect novel deletions at the respective loci.

2.2.10. Methods for Detection of α-Globin Gene Triplications and Quadruplications

Traditionally, Southern blot is required to detect both anti-3.7 and anti-4.2 triplication and quadruplication alleles [65]. Today, PCR methods are available for detection of the anti-3.7 and anti-4.2 alleles [22, 23] [66]. In the anti-3.7/4.2 multiplex-PCR assay, the primers were designed to only amplify either an anti-3.7 or an anti-4.2 junction fragment from the respective anti-3.7 or anti-4.2 allele (Figure 10).

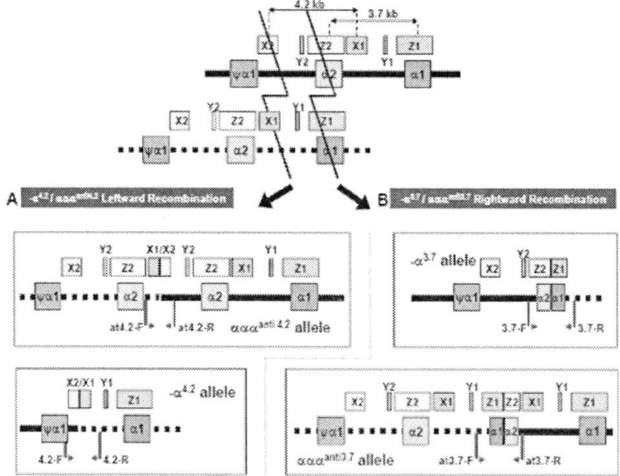

Figure 10. Origin of the α-globin single gene deletions and their corresponding gene triplications and PCR primers used to detect their junction fragments. (A) Crossovers between misaligned X boxes give rise to -α$^{4.2}$ and anti-4.2 chromosomes (leftward crossover). (B) Crossovers between misaligned Z boxes give rise to -α$^{3.7}$ and anti-3.7 chromosomes (rightward crossover). For further details, please refer to [66].

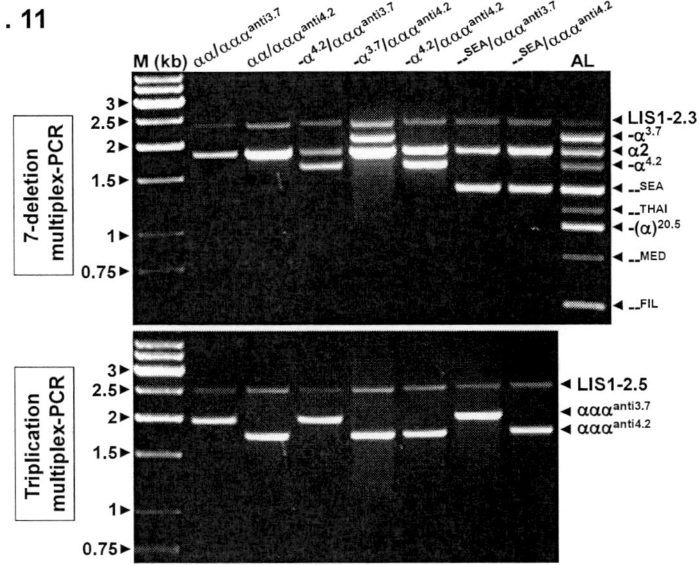

Figure 11. PCR assays used to determine α-globin genotype. (A) 7-deletion multiplex-PCR; (B) Anti-3.7/4.2 multiplex-PCR.

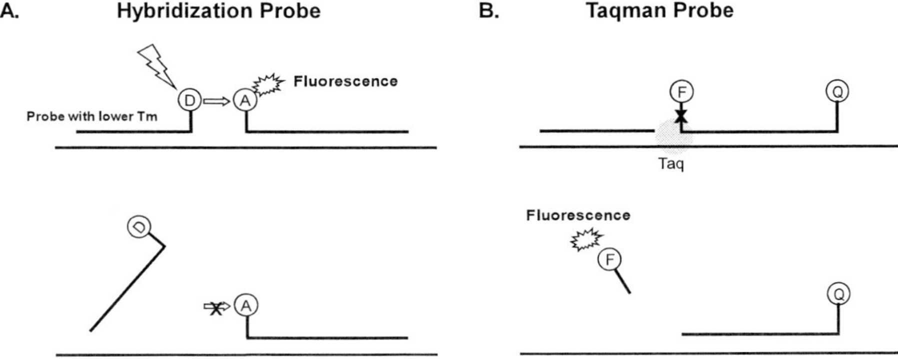

Figure 12. Hybridization Probe and Taqman probe detection technology. (A) Hybridization probes are paired and labeled with a donor fluorophore (D) on one probe and an acceptor fluorophore (A) on the other. Excitation of the donor fluorophore results in fluorescence emission. Close proximity of annealed probes facilitates the excitation of the acceptor fluorophore via fluorescence resonance energy transfer (FRET) and fluorescence emission at a different wavelength. Temperature mediated dissociation of either or both probes results in loss of fluorescence. (B) Taqman probes are dually labeled with a fluorescent label (F) at the 5′ end and a fluorescent quencher (Q) at the 3′ end. During the extension phase of a PCR reaction, Taq polymerase excises the fluorescently labeled 5′ nucleotide off the probe. The released fluorescent moiety emits fluorescence as it is no longer in close proximity to the quencher.

No amplification product would be generated from a normal (αα) or deletional (−α or −−) allele. The different amplicon sizes of the anti-3.7 and anti-4.2 junction fragments (1.9 and 1.7 kb respectively) allow for easy identification by agarose gel electrophoresis (Figure 11). Recently, MLPA has also been applied for the detection of α-globin gene triplications and other re-arrangements [67, 68].

2.2.11. Real-Time PCR

Real-time PCR allows real-time measurement of fluorescence in the early exponential phase of the amplification reaction, thus enabling precise quantification. In addition, post-PCR melting curve analysis is a powerful too to characterize the end-point PCR product. Several other advantages of real-time PCR analysis over traditional PCR include elimination of post-PCR processing steps and a wide dynamic range of detection with a high degree of sensitivity. These advantages make real-time PCR an excellent tool for copy number analysis and genotyping.

Quantitative real time PCR has been applied for diagnosis of α-thalassemia [69-71]. During real time PCR, the progress of amplification can be monitored through the fluorescence of double-stranded intercalating dyes (SYBR Green I or SYTO9), or fluorescence labeled allele specific probes, and normalized against an internal reference sequence. By comparing the wild type alleles and deletion alleles, sample genotype (heterozygous, homozygous or wild-type) can be assigned. Instead of quantitating mutant relative to normal alleles, Timmann et al. [72] determined the ratio of α2/α1 globin genes by end-point melting curve analysis. The α1 and α2 globin genes were amplified together using a primer pair annealing to sequence identical between both genes, and were then differentiated and quantitated by analysis of melting peaks generated from melting of a hybridization probe that is perfectly matched to α2, but contains mismatches to α1. Post-PCR melting curve analysis can also be used to detect deletion alleles and the wild type alleles of α-thalassemia [73, 74]. In this assay, the PCR fragments of wild type and deletion alleles are designed to melt at different temperatures, thus generating distinct and clearly differentiated melting peaks.

Real-time PCR can also be used to screen for point mutations in β-thalassemia, by using fluorescence labeled hybridization probes or Taqman probes that cover the regions containing the mutations [75] (Figure 12).

The presence of a mutation in the amplified sample DNA leads to a probe-template mismatch, which translates into a melting temperature (Tm) shift of 5–10 °C, thus allowing easy distinction between wild-type and mutant alleles, and determination of disease genotype. The use of two or more colored probes allows more than one mutation to be screened in a single reaction. This method has been successfully applied to screen for a spectrum of β-thalassemia mutations in the Greek population for application to carrier screening and prenatal diagnosis [76] and for the prenatal diagnosis of sickle cell disease and detection of maternal contamination [77].

2.2.12. High Resolution Melting Analysis

High resolution melting (HRM) analysis is a new post-PCR mutation scanning tool that identifies the presence of sequence alterations without requiring any manual post-PCR treatments, and is a variation of melting curve analysis. PCR amplicons are slowly heated until fully denatured while the fluorescence is monitored in real-time [78]. Amplicons of a sequence containing a mutation generate altered melting curves compared with wild type samples. By determining the difference in fluorescence between the test sample and a wild-type control sample at each temperature point, a difference plot is generated. Any deviation from the no difference flat line is indicative of a sequence difference from wild-type sequence. HRM has a very high sensitivity for detecting changes down to the single nucleotide level when using saturating dyes such as LC Green, EVA Green or SYTO 9 [79], and has been used for detection of non-deletional α-thalassemia mutations, such as Hb Constant Spring, Hb Quong Sze and Hb Westmead [80]. It has also been used for screening β globin point mutations, such as c.−78A>G, c.−79A>G, c.2T>G, c.79_80insT, c.84_85insC, c.123_124insT, c.125_128delTCTT, c.130 G>T, c.170G>A, c.216_217ins A and c.316–197 C>T [81].

2.2.13. Microarray: High-Throughput Approaches in Molecular Diagnostics

Normally, separate tests are required for detection of α-thalassemia (deletional mutations) and β-thalassemia (non-deletional mutations). Microarray technology provides a solution to simultaneously screen for multiple deletional and non-deletional mutations on a single platform [82, 83]. For the microarray described in these reports, a pair of probes is used for each point mutation, one complementary to the wild type sequence and the other to the mutant. For deletional mutations, probes are designed to detect the presence of a gap-PCR junction fragment. Sample DNA is prepared by a multiplex PCR reaction which simultaneously amplifies the β-globin gene fragment containing all the expected point mutations, as well as the wild-type and deletion junction fragments at the α-globin gene cluster. Amplified products are labeled and hybridized onto the microarray, followed by fluorescence scanning. A quantitative analysis is then performed by calculating the relative signal ratio between the wild type probe and mutant probe (W/M). A signal threshold is established for categorizing positive and negative signals, since some background hybridization normally occurs especially when mutant and wildtype probes differ by only one or two nucleotides.

Another type of microarray, arrayed primer extension (APEX) technology, has been used for β-thalassemia [84-86]. This technique adopts a similar genotyping strategy as minisequencing. Built on the chip platform, APEX is a solid-phase minisequencing assay in which the detection primers are immobilized on a glass slide.

3. Prenatal and Preimplantation Genetic Diagnosis of Thalassemia

Prenatal and preimplantation genetic diagnoses (PGD) of thalassemia are both aimed at reducing the incidence of thalassemia and the number of affected individuals in high risk populations. Theoretically, all of the DNA diagnostic methods reported for thalassemia diagnosis are suitable for prenatal diagnosis. However, due to the possibility of contamination of the fetal sample with maternal tissue during fetal sampling, an additional test to rule out maternal DNA contamination is required. This is usually accomplished through the PCR analysis of highly polymorphic microsatellite markers [87].

Elective pregnancy termination after obtaining positive results from prenatal diagnosis is currently the only option to avoid delivering an affected child, and can be emotionally traumatic for the coupled involved. PGD represents an alternative option that avoids this dilemma. In PGD, 1-2 blastomeres are removed from a 3 day old embryo for genetic testing to rule out a specific genetic disorder, so that only those embryos that are clinically unaffected are selected for uterine transfer.

PGD for thalassemia has become one of the most common procedures performed. Until the end of 2006, almost 430 PGD cycles for thalassemia and sickle cell syndromes have been reported [88]. The first cycles of PGD for thalassemia was reported in 1998, which applied nested-PCR with restriction enzyme analysis of polar bodies [89]. Other approaches used either DGGE analysis [90] or reverse dot-blot to screen for β-thalassemia [91]. Several other protocols employ restriction enzyme digestion to screen for β-thalassemia [92], SSCP analysis for PGD of β-thalassemia [93], and even fluorescent gap-PCR [94]. Fluorescent gap-PCR assays that include linked polymorphic markers [95] greatly reduce the misdiagnosis rate and are useful in detecting DNA contamination. A PGD assay is now available that is applicable to all common deletional determinants of Hb Bart's syndrome [96]. In this assay, microsatellite markers 16PTEL05 and 16PTEL06, which lie within the deletional region of the α-globin gene cluster were examined to screen for Hb Bart's syndrome affected embryos (Figure 2).

Other recent methods which have been reported for PGD of thalassemia include minisequencing or real-time PCR. Minisequencing is ideally suited for single cell analysis as it requires only short DNA fragments, which are much more efficiently amplified than larger fragments [97]. Real-time PCR with hybridization probes also provides a rapid and accurate approach for single cell genotyping, and this method has been used for PGD of β-thalassemia. Since most mutations are clustered in a relatively small region of the β-globin gene, multiplex genotyping by hybridization probes is possible [98].

Conclusion

The compactness of the globin genes, and the innovative adaptation of PCR-based strategies, has resulted in the molecular diagnosis of the thalassemias being predominantly PCR-based and no longer reliant on Southern blotting. However, molecular diagnostic strategies are constantly improving to make them simpler, cheaper, and more universal. Higher throughout approaches will be required to meet increasing demands for newborn and carrier screening. Increasingly, automation of sample preparation, assay setup, and mutation detection will feature more prominently in molecular diagnostics in the future, and microarray and next generation sequencing platforms will likely play a pivotal role. For lower throughput requirements such as prenatal and pre-implantation diagnosis, real-time PCR and capillary electrophoresis platforms, with their high detection sensitivity and specificity and simplified DNA analysis, are expected to remain key diagnostic platforms.

References

[1]   Weatherall, DJ; Clegg, JB. Inherited hemoglobin disorders: an increasing global health problem. *Bull World Health Organ*, 2001, 79, 704-712.

[2]   Lorey, F. Asian immigration and public health in California: thalassemia in newborns in California. *J. Pediatr. Hematol. Oncol.*, 2000, 22, 564-566.

[3]   Vichinsky, EP. Changing patterns of thalassemia worldwide. *Ann. N. Y. Acad. Sci.*, 2005, 1054, 18-24.

[4]   Henderson, S; Timbs, A, et al. Incidence of haemoglobinopathies in various populations - the impact of immigration. *Clin. Biochem.*, 2009, 42, 1745-1756.

[5]   Weatherall, DJ. Genetic variation and susceptibility to infection: the red cell and malaria. *Br. J. Haematol.*, 2008, 141, 276-286.

[6]   Weatherall, DJ; Clegg, JB. The thalassaemia syndromes. 4th ed. Oxford: Blackwell Science; 2001.

[7]   Higgs, DR; Wood, WG, et al. A major positive regulatory region located far upstream of the human alpha-globin gene locus. *Genes. Dev.*, 1990, 4, 1588-1601.

[8]   Liebhaber, SA; Griese, EU, et al. Inactivation of human alpha-globin gene expression by a de novo deletion located upstream of the alpha-globin gene cluster. *Proc. Natl. Acad. Sci. USA*, 1990, 87, 9431-9435.

[9]   Hughes, JR; Cheng, JF, et al. Annotation of cis-regulatory elements by identification, subclassification, and functional assessment of multispecies conserved sequences. *Proc. Natl. Acad. Sci. USA*, 2005, 102, 9830-9835.

[10]   Goh, SH; Lee, YT, et al. A newly discovered human alpha-globin gene. *Blood,* 2005, 106, 1466-1472.

[11]   Weatherall, DJ, The thalassemias, In: The molecular basis of blood diseases, Stamatoyannopoulos, G, Majerus, PW, et al., Editors. 2001, Philadelphia: W.B. Saunders. 183-226.

[12]   Patrinos, GP; Giardine, B, et al. Improvements in the HbVar database of human hemoglobin variants and thalassemia mutations for population and sequence variation studies. *Nucleic Acids Res.*, 2004, 32, D537-541.

[13]   Giardine, B; van Baal, S, et al. HbVar database of human hemoglobin variants and thalassemia mutations: 2007 update. *Hum. Mutat.*, 2007, 28, 206.

[14]   Chui, DH; Fucharoen, S, et al. Hemoglobin H disease: not necessarily a benign disorder. *Blood,* 2003, 101, 791-800.

[15]   Huisman, THJ; Carver, MFH, et al. A Syllabus of Thalassemia Mutations. Augusta, GA: The Sickle Cell Anemia Foundation; 1997.

[16]   Grosveld, F; van Assendelft, GB, et al. Position-independent, high-level expression of the human beta-globin gene in transgenic mice. *Cell*, 1987, 51, 975-985.

[17]   Clarke, GM; Higgins, TN. Laboratory investigation of hemoglobinopathies and thalassemias: review and update. *Clin. Chem.*, 2000, 46, 1284-1290.

[18]   Galanello, R; Origa, R. Beta-thalassemia. *Orphanet J. Rare Dis.*, 2010, 5, 11.

[19]   Bain, BJ. Haemoglobinopathy Diagnosis: Blackwell Science Ltd; 2001.

[20]   The laboratory diagnosis of haemoglobinopathies. *Br. J. Haematol.*, 1998, 101, 783-792.

[21]   Fischel-Ghodsian, N; Vickers, MA, et al. Characterization of two deletions that remove the entire human zeta-alpha globin gene complex (- -THAI and - -FIL). *Br. J. Haematol.*, 1988, 70, 233-238.

[22]   Dode, C; Krishnamoorthy, R, et al. Rapid analysis of -alpha 3.7 thalassaemia and alpha alpha alpha anti 3.7 triplication by enzymatic amplification analysis. *Br. J. Haematol.*, 1993, 83, 105-111.

[23]   Liu, YT; Old, JM, et al. Rapid detection of alpha-thalassaemia deletions and alpha-globin gene triplication by multiplex polymerase chain reactions. *Br. J. Haematol.*, 2000, 108, 295-299.

[24] Chong, SS; Boehm, CD, et al. Single-tube multiplex-PCR screen for common deletional determinants of alpha-thalassemia. *Blood*, 2000, 95, 360-362.

[25] Chong, SS; Boehm, CD, et al. Simplified multiplex-PCR diagnosis of common Southeast Asian deletional determinants of alpha-thalassemia [In Process Citation]. *Clin. Chem.*, 2000, 46, 1692-1695.

[26] Baysal, E; Huisman, TH. Detection of common deletional alpha-thalassemia-2 determinants by PCR. *Am. J. Hematol.*, 1994, 46, 208-213.

[27] Eng, B; Patterson, M, et al. PCR-based diagnosis of the Filipino (--(FIL)) and Thai (--(THAI)) alpha- thalassemia-1 deletions. *Am. J. Hematol.*, 2000, 63, 54-56.

[28] Tan, AS; Quah, TC, et al. A rapid and reliable 7-deletion multiplex polymerase chain reaction assay for alpha-thalassemia. *Blood*, 2001, 98, 250-251.

[29] Bergstrome, JA; Poon, A. Evaluation of a single-tube multiplex polymerase chain reaction screen for detection of common alpha-thalassemia genotypes in a clinical laboratory. *Am. J. Clin. Pathol.*, 2002, 118, 18-24.

[30] Vandenplas, S; Higgs, DR, et al. Characterization of a new alpha zero thalassaemia defect in the South African population. *Br. J. Haematol.*, 1987, 66, 539-542.

[31] Shaji, RV; Eunice, SE, et al. Determination of the breakpoint and molecular diagnosis of a common alpha-thalassaemia-1 deletion in the Indian population. *Br. J. Haematol.*, 2003, 123, 942-947.

[32] Old, JM; Varawalla, NY, et al. Rapid detection and prenatal diagnosis of beta-thalassaemia: studies in Indian and Cypriot populations in the UK. *Lancet*, 1990, 336, 834-837.

[33] Rady, MS; Baffico, M, et al. Identification of Mediterranean beta-thalassemia mutations by reverse dot-blot in Italians and Egyptians. *Hemoglobin*, 1997, 21, 59-69.

[34] Sutcharitchan, P; Saiki, R, et al. Reverse dot-blot detection of the African-American beta-thalassemia mutations. *Blood*, 1995, 86, 1580-1585.

[35] Sutcharitchan, P; Saiki, R, et al. Reverse dot-blot detection of Thai beta-thalassaemia mutations. *Br. J. Haematol.*, 1995, 90, 809-816.

[36] Winichagoon, P; Saechan, V, et al. Prenatal diagnosis of beta-thalassaemia by reverse dot-blot hybridization. *Prenat. Diagn*, 1999, 19, 428-435.

[37] Ng, IS; Ong, JB, et al. Beta-thalassemia mutations in Singapore--a strategy for prenatal diagnosis. *Hum Genet*, 1994, 94, 385-388.

[38] Chan, V; Yam, I, et al. A reverse dot-blot method for rapid detection of non-deletion alpha thalassaemia. *Br. J. Haematol.*, 1999, 104, 513-515.

[39] Lindeman, R; Hu, SP, et al. Polymerase chain reaction (PCR) mutagenesis enabling rapid non-radioactive detection of common beta-thalassaemia mutations in Mediterraneans. *Br. J. Haematol.*, 1991, 78, 100-104.

[40] Gorakshakar, AC; Pawar, AR, et al. Potential of denaturing gradient gel electrophoresis for scanning of beta-thalassemia mutations in India. *Am. J. Hematol.*, 1999, 61, 120-125.

[41] Dozy, AM; Kan, YW. Characterization of beta-thalassemia mutations by denaturing gradient gel electrophoresis: patterns in the Mediterranean mutations. *Clin. Genet*, 1994, 45, 221-227.

[42] Losekoot, M; Fodde, R, et al. Denaturing gradient gel electrophoresis and direct sequencing of PCR amplified genomic DNA: a rapid and reliable diagnostic approach to beta thalassaemia. *Br. J. Haematol.*, 1990, 76, 269-274.

[43] Cai, S; Kan, Y. Identification of the multiple beta-thalassemia mutations by denaturing gradient gel electrophoresis. *J. Clin. Invest.*, 1990, 85, 550-553.

[44] Harteveld, KL; Heister, AJ, et al. Rapid detection of point mutations and polymorphisms of the alpha- globin genes by DGGE and SSCA. *Hum. Mutat.*, 1996, 7, 114-122.

[45] Ahmed, S; Saleem, M, et al. Prenatal diagnosis of beta-thalassaemia in Pakistan: experience in a Muslim country. *Prenat. Diagn*, 2000, 20, 378-383.

[46] Tan, JA; Tay, JS, et al. The amplification refractory mutation system (ARMS): a rapid and direct prenatal diagnostic technique for beta-thalassaemia in Singapore. *Prenat. Diagn*, 1994, 14, 1077-1082.

[47] Fortina, P; Dotti, G, et al. Detection of the most common mutations causing beta-thalassemia in Mediterraneans using a multiplex amplification refractory mutation system (MARMS). *PCR Methods Appl.*, 1992, 2, 163-166.

[48] Tan, KL; Tan, JA, et al. Combine-ARMS: a rapid and cost-effective protocol for molecular characterization of beta-thalassemia in Malaysia. *Genet. Test*, 2001, 5, 17-22.

[49] Eng, B; Patterson, M, et al. Detection of severe nondeletional alpha-thalassemia mutations using a single-tube multiplex ARMS assay. *Genet. Test*, 2001, 5, 327-329.

[50] Newton, CR; Graham, A, et al. Analysis of any point mutation in DNA. The amplification refractory mutation system (ARMS). *Nucleic Acids Res.*, 1989, 17, 2503-2516.

[51] Dode, C; Rochette, J, et al. Locus assignment of human alpha globin mutations by selective amplification and direct sequencing. *Br. J. Haematol.*, 1990, 76, 275-281.

[52] Molchanova, TP; Pobedimskaya, DD, et al. A simplified procedure for sequencing amplified DNA containing the alpha 2- or alpha 1-globin gene. *Hemoglobin*, 1994, 18, 251-255.

[53] Thong, MK; Law, HY, et al. Molecular heterogeneity of beta-thalassaemia in Malaysia: a practical approach to diagnosis. *Ann. Acad. Med. Singapore*, 1996, 25, 79-83.

[54] Metzker, ML. Sequencing technologies - the next generation. *Nat. Rev. Genet.*, 2010, 11, 31-46.

[55] Haywood, A; Dreau, H, et al. Screening for clinically significant non-deletional alpha thalassaemia mutations by pyrosequencing. *Ann. Hematol.*, 2010, 89, 1215-1221.

[56] Salk, JJ; Sanchez, JA, et al. Direct amplification of single-stranded DNA for pyrosequencing using linear-after-the-exponential (LATE)-PCR. *Anal. Biochem.*, 2006, 353, 124-132.

[57] Makridakis, NM; Reichardt, JK. Multiplex automated primer extension analysis: simultaneous genotyping of several polymorphisms. *Biotechniques,* 2001, 31, 1374-1380.

[58] Wang, W; Kham, SK, et al. Multiplex minisequencing screen for common Southeast Asian and Indian beta-thalassemia mutations. *Clin. Chem.*, 2003, 49, 209-218.

[59] Wang, W; Ma, ES, et al. Multiple minisequencing screen for seven southeast Asian nondeletional alpha-thalassemia mutations. *Clin. Chem.*, 2003, 49, 800-803.

[60] Schouten, JP; McElgunn, CJ, et al. Relative quantification of 40 nucleic acid sequences by multiplex ligation-dependent probe amplification. *Nucleic Acids Res*, 2002, 30, e57.

[61] Gallienne, AE; Dreau, HM, et al. Multiplex ligation-dependent probe amplification identification of 17 different beta-globin gene deletions (including four novel mutations) in the UK population. *Hemoglobin*, 2009, 33, 406-416.

[62] Lee, ST; Yoo, EH, et al. Multiplex ligation-dependent probe amplification screening of isolated increased HbF levels revealed three cases of novel rearrangements/deletions in the beta-globin gene cluster. *Br. J. Haematol.*, 2010, 148, 154-160.

[63] Liu, JZ; Han, H, et al. Detection of alpha-thalassemia in China by using multiplex ligation-dependent probe amplification. *Hemoglobin*, 2008, 32, 561-571.

[64] So, CC; So, AC, et al. Detection and characterisation of beta-globin gene cluster deletions in Chinese using multiplex ligation-dependent probe amplification. *J. Clin. Pathol.*, 2009, 62, 1107-1111.

[65] Gu, YC; Landman, H, et al. Two different quadruplicated alpha globin gene arrangements. *Br. J. Haematol.*, 1987, 66, 245-250.

[66] Wang, W; Ma, ES, et al. Single-tube multiplex-PCR screen for anti-3.7 and anti-4.2 alpha-globin gene triplications. *Clin. Chem.*, 2003, 49, 1679-1682.

[67] Harteveld, CL; Refaldi, C, et al. Segmental duplications involving the alpha-globin gene cluster are causing beta-thalassemia intermedia phenotypes in beta-thalassemia heterozygous patients. *Blood Cells Mol. Dis.*, 2008, 40, 312-316.

[68] Colosimo, A; Gatta, V, et al. Application of MLPA assay to characterize unsolved alpha-globin gene rearrangements. *Blood Cells Mol. Dis.*, 2011, 46, 139-144.

[69] Sun, CF; Lee, CH, et al. Real-time quantitative PCR analysis for alpha-thalassemia-1 of Southeast Asian type deletion in Taiwan. *Clin. Genet.*, 2001, 60, 305-309.

[70] Pornprasert, S; Sukunthamala, K, et al. Analysis of real-time SYBR-polymerase chain reaction cycle threshold for diagnosis of the alpha-thalassemia-1 Southeast Asian type deletion: application to carrier screening and prenatal diagnosis of Hb Bart's hydrops fetalis. *Hemoglobin*, 2008, 32, 393-402.

[71] Chan, V; Yip, B, et al. Quantitative polymerase chain reaction for the rapid prenatal diagnosis of homozygous alpha-thalassaemia (Hb Barts hydrops fetalis). *Br. J. Haematol.*, 2001, 115, 341-346.

[72] Timmann, C; Moenkemeyer, F, et al. Diagnosis of alpha+-thalassemias by determining the ratio of the two alpha-globin gene copies by oligonucleotide hybridization and melting curve analysis. *Clin. Chem.*, 2005, 51, 1711-1713.

[73] Liu, J; Tang, N, et al. Improvement in the detection of alpha0- and deletional alpha-thalassemia by real-time PCR combined with dissociation curve analysis. *Acta Haematol.*, 2009, 122, 17-22.

[74] Liu, JZ; Yan, M, et al. Molecular prenatal diagnosis of alpha-thalassemia using real-time and multiplex polymerase chain reaction methods. *Hemoglobin*, 2008, 32, 553-560.

[75] Wilhelm, J; Pingoud, A. Real-time polymerase chain reaction. *Chembiochem.*, 2003, 4, 1120-1128.

[76] Vrettou, C; Traeger-Synodinos, J, et al. Rapid screening of multiple beta-globin gene mutations by real-time PCR on the LightCycler: application to carrier screening and prenatal diagnosis of thalassemia syndromes. *Clin. Chem.*, 2003, 49, 769-776.

[77] Costa, C; Pissard, S, et al. A one-step real-time PCR assay for rapid prenatal diagnosis of sickle cell disease and detection of maternal contamination. *Mol. Diagn*, 2003, 7, 45-48.

[78] Montgomery, J; Wittwer, CT, et al. Simultaneous mutation scanning and genotyping by high-resolution DNA melting analysis. *Nat. Protoc.*, 2007, 2, 59-66.

[79] Wittwer, CT; Reed, GH, et al. High-resolution genotyping by amplicon melting analysis using LCGreen. *Clin. Chem.*, 2003, 49, 853-860.

[80] Li, R; Liao, C, et al. High-resolution melting analysis of the three common nondeletional alpha-thalassemia mutations in the Chinese population: Hbs Constant Spring, Quong Sze and Westmead. *Hemoglobin*, 2010, 34, 587-593.

[81] Shih, HC; Er, TK, et al. Rapid identification of HBB gene mutations by high-resolution melting analysis. *Clin Biochem*, 2009, 42, 1667-1676.

[82] Ye, BC; Liu, H, et al. Simultaneous detection of alpha-thalassemia and beta-thalassemia by oligonucleotide microarray. *Haematologica*, 2004, 89, 1010-1012.

[83] Ye, BC; Zhang, Z, et al. Molecular analysis of alpha/beta-thalassemia in a southern Chinese population. *Genet. Test*, 2007, 11, 75-83.

[84] Lu, Y; Kham, SK, et al. Arrayed primer extension: a robust and reliable genotyping platform for the diagnosis of single gene disorders: beta-thalassemia and thiopurine methyltransferase deficiency. *Genet. Test*, 2005, 9, 212-219.

[85] van Moorsel, CH; van Wijngaarden, EE, et al. beta-Globin mutation detection by tagged single-base extension and hybridization to universal glass and flow-through microarrays. *Eur. J. Hum. Genet.*, 2004, 12, 567-573.

[86] Gemignani, F; Perra, C, et al. Reliable detection of beta-thalassemia and G6PD mutations by a DNA microarray. *Clin. Chem.*, 2002, 48, 2051-2054.

[87] Decorte, R; Cuppens, H, et al. Rapid detection of hypervariable regions by the polymerase chain reaction technique. *DNA Cell Biol.*, 1990, 9, 461-469.

[88] Goossens, V; Harton, G, et al. ESHRE PGD Consortium data collection IX: cycles from January to December 2006 with pregnancy follow-up to October 2007. *Hum. Reprod.*, 2009, 24, 1786-1810.

[89] Kuliev, A; Rechitsky, S, et al. Preimplantation diagnosis of thalassemias. *J. Assist. Reprod. Genet.*, 1998, 15, 219-225.

[90] Vrettou, C; Palmer, G, et al. A widely applicable strategy for single cell genotyping of beta-thalassaemia mutations using DGGE analysis: application to preimplantation genetic diagnosis. *Prenat. Diagn*, 1999, 19, 1209-1216.

[91] Jiao, Z; Zhou, C, et al. Birth of healthy children after preimplantation diagnosis of beta-thalassemia by whole-genome amplification. *Prenat. Diagn*, 2003, 23, 646-651.

[92] Kuliev, A; Rechitsky, S, et al. Birth of healthy children after preimplantation diagnosis of thalassemias. *J. Assist. Reprod. Genet.*, 1999, 16, 207-211.

[93] el-Hashemite, N; Wells, D, et al. Single cell detection of beta-thalassaemia mutations using silver stained SSCP analysis: an application for preimplantation diagnosis. *Mol. Hum. Reprod.*, 1997, 3, 693-698.

[94] De Rycke, M; Van de Velde, H, et al. Preimplantation genetic diagnosis for sickle-cell anemia and for beta-thalassemia. *Prenat. Diagn*, 2001, 21, 214-222.

[95] Piyamongkol, W; Harper, JC, et al. Preimplantation genetic diagnostic protocols for alpha- and beta-thalassaemias using multiplex fluorescent PCR. *Prenat. Diagn*, 2001, 21, 753-759.

[96] Wang, W; Yap, CH, et al. Simplified PGD of common determinants of hemoglobin Bart's hydrops fetalis syndrome using multiplex-microsatellite PCR. *Reprod. Biomed. Online*, 2010, 21, 642-648.

[97] Fiorentino, F; Magli, MC, et al. The minisequencing method: an alternative strategy for preimplantation genetic diagnosis of single gene disorders. *Mol. Hum. Reprod.*, 2003, 9, 399-410.

[98] Vrettou, C; Traeger-Synodinos, J, et al. Real-time PCR for single-cell genotyping in sickle cell and thalassemia syndromes as a rapid, accurate, reliable, and widely applicable protocol for preimplantation genetic diagnosis. *Hum. Mutat.*, 2004, 23, 513-521.

In: Genetic Diagnoses
Editor: R. J. Sarma, pp. 53-97

ISBN: 978-1-61324-866-9
© 2012 Nova Science Publishers, Inc.

Chapter III

Fuzzy Logic and the Least Squares Method in Diagnosis Problem Solving

Alexander P. Rotshtein[*] and Hanna B. Rakytyanska
[1]Industrial Engineering and Management Dept.
Jerusalem College of Technology – Machon Lev, Jerusalem, Israel
[2]Soft Ware Design Dept. Vinnitsa National Technical University
Vinnitsa, Ukraine

Abstract

This chapter deals with restoration and identification of the causes (diagnoses) through the observed effects (symptoms) on the basis of fuzzy relational calculus and the least squares method. Fuzzy relational calculus plays the central role as a uniform platform for inverse problem resolution on various fuzzy approximation operators. The diagnosis problem, which is based on a cause and effect analysis and abductive reasoning, can be formally described by fuzzy relations and fuzzy IF-THEN rules. In this chapter, we propose an approach for building fuzzy systems of diagnosis, which enables solving fuzzy relational equations together with design and tuning of multiple variable linguistic models on the basis of expert and experimental information. We suggest some procedures of numerical solution of the fuzzy relational equations using genetic algorithms. The procedures envisage the optimal solution growing from a set of primary variants by extending the simplified solution set to the case of multidimensional fuzzy relational equations. The essence of tuning consists of the selection of such membership functions of the fuzzy terms for the input and output variables (causes and effects) and such «causes-effects» fuzzy relations, which provide minimal difference between theoretical and experimental results of diagnosis. The problem of tuning is formulated as an optimization problem, which is solved by using a genetic algorithm. The efficiency of the proposed models and algorithms is illustrated by the computer experiments and the real examples from technical diagnosis.

---

[*] e-mail: rot@jct.ac.il.

1. Introduction

Fuzzy logic is a handy tool for expert information formalization while simulating cause-effect connections in diagnostic problems [1]. Two main types of structures, built with the aid of fuzzy IF-THEN rules and fuzzy relations, are used in the case of diagnosis of problems where physical mechanisms are not well known due to high complexity and nonlinearity [2]. This chapter deals with restoration and identification of the causes (diagnoses) through the observed effects (symptoms) on the basis of fuzzy relational calculus and the least squares method.

Fuzzy relational calculus plays the central role as a uniform platform for inverse problem resolution on various fuzzy approximation operators [3]. A model for diagnosis can be built on the basis of Zadeh's compositional rule of inference, in which the fuzzy matrix of «causes-effects» relations serves as the support of the diagnostic information [1]. In relational models, inputs-outputs connection is described by a system of multidimensional fuzzy relational equations with extended max-min law of composition [2]. In rule-based models, inputs-outputs connection is described by a hierarchical system of simplified fuzzy relational equations with max-min and dual min-max law of composition [4, 5]. The problem of inputs restoration and identification is formulated in the form of inverse fuzzy logical inference and requires solution of a system of fuzzy relational equations. In this case some renewal of causes takes place according to observable effects.

Inverse problem resolution is of interest to both relational models and rule-based ones. In the case of multiple variable linguistic model, the cause-effect dependency is extended to the multidimensional fuzzy relational structure [2]. The insufficient use of the inverse logical inference is stipulated through the lack of effective algorithms for solving multidimensional fuzzy relational equations. The solution existence problem and solvability criteria for simplified fuzzy relation equations have been studied for many years [3, 6 – 12]. The system of such equations can be considered as a single input single output fuzzy approximator [13, 14]. While the theoretical foundations of simplified fuzzy relational equations are well developed they call for more efficient use of their potential in system modelling [3]. The problem consists not only in solving the system of fuzzy logical equations, but also in construction of such a fuzzy relational model, which provides maximal proximity between model and real results of diagnosis.

In those cases, when domain experts are involved in developing fuzzy models, construction of the cause-effect connections can be considered as rough tuning of the fuzzy relational model [15]. The observed (output) and diagnosed (input) parameters of a system are considered as linguistic variables [1]. Fuzzy terms, e.g., «temperature rise», «pressure drop» etc., associated with causes and effects are used for these variables evaluation. The use of the expert relational matrix cannot guarantee the coincidence of theoretical results of diagnosis and real data. In other words, the «quality» of the model strongly depends on the «quality» of the expert forming the diagnostic matrix. In addition, the problem of solving fuzzy relational equations is still relevant – as of yet there does not exist a satisfactory answer for computing a complete solution set [3].

In works [16 – 18], an expert system using a genetic algorithm [19, 20] as a tool to solve the diagnosis problem was proposed. The diagnosis problem based on a cause and effect analysis was formally described by the single input single output fuzzy relation approximator.

The suggested genetically based approach consists of formulating and solving the optimization problems, which find the roots of fuzzy relation equations and tune the fuzzy relation matrix using the readily available experimental data.

In this chapter, we propose an approach for building fuzzy systems of diagnosis, which enables solving fuzzy relational equations together with design and tuning of multiple variable linguistic models on the basis of expert and experimental information.

The problem of diagnosis amounts to solving a system of fuzzy relational equations on various fuzzy approximation operators and with different kinds of composition laws for fuzzy relations by using the least squares method. The search for a system solution requires resolution for the optimization problem. We suggest some procedures of numerical solution of the fuzzy relational equations using genetic algorithms. The procedures envisage the optimal solution growing from a set of primary variants by extending the simplified solution set to the case of multidimensional fuzzy relational equations.

The essence of tuning consists of the selection of such membership functions of the fuzzy terms for the input and output variables (causes and effects) and such «causes-effects» fuzzy relations, which provide minimal difference between theoretical and experimental results of diagnosis. The problem of tuning is formulated as an optimization problem, which is solved by using a genetic algorithm.

The efficiency of the proposed models and algorithms is illustrated by the computer experiments and the real examples from technical diagnosis.

2. Diagnosis Problem Statement

The diagnosis object is treated as a black box with n inputs and m outputs (Figure 1). Outputs of the object are associated with the observed effects (symptoms). Inputs correspond to the causes of the observed effects (diagnoses). The problem of diagnosis consists of restoration and identification of the causes (inputs) through the observed effects (outputs). Inputs and outputs can be considered as linguistic variables given on the corresponding universal sets. Fuzzy terms are used for these linguistic variables evaluation. We shall denote the following:

$\{x_1, x_2, ..., x_n\}$ is the set of input parameters, $x_i \in [\underline{x}_i, \overline{x}_i]$, $i = \overline{1,n}$;

$\{y_1, y_2, ..., y_m\}$ is the set of output parameters, $y_j \in [\underline{y}_j, \overline{y}_j]$, $j = \overline{1,m}$;

$\{c_{i1}, c_{i2}, ..., c_{ik_i}\}$ is the set of linguistic terms for parameter x_i evaluation, $i = \overline{1,n}$;

$\{e_{j1}, e_{j2}, ..., e_{jq_j}\}$ is the set of linguistic terms for parameter y_j evaluation, $j = \overline{1,m}$.

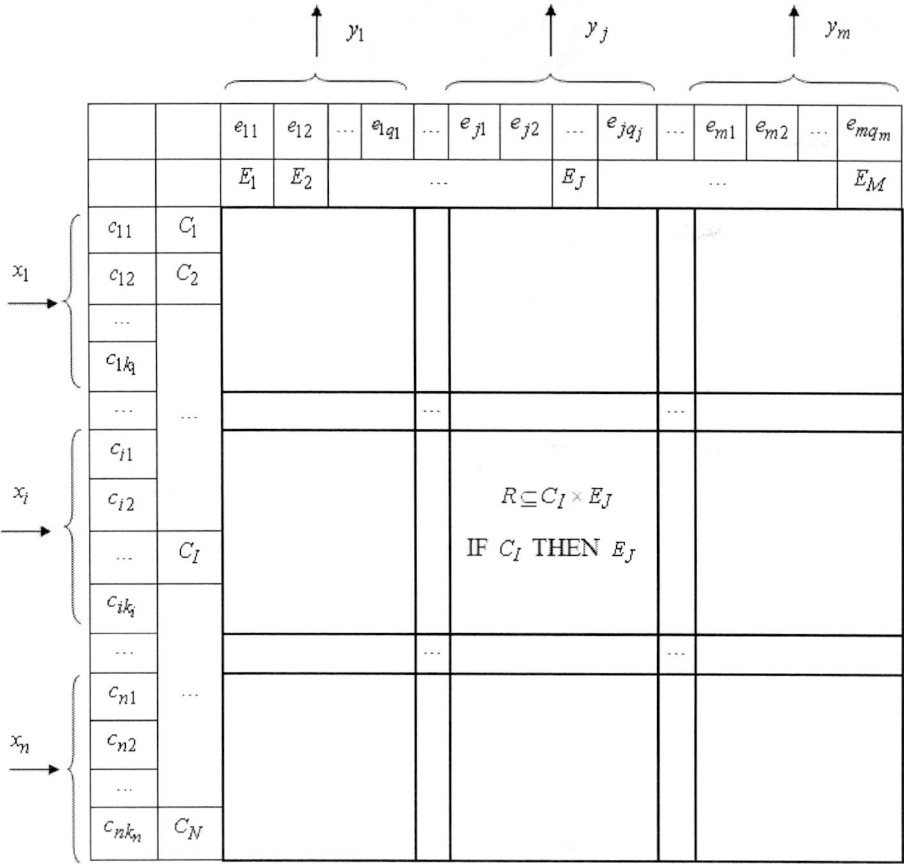

Figure 1. Object of diagnosis.

Each term-assessment is described with the help of a fuzzy set:

$$c_{il} = \{(x_i, \mu^{c_{il}}(x_i))\}, \ i = \overline{1,n}, \ l = \overline{1,k_i};$$

$$e_{jp} = \{(y_j, \mu^{e_{jp}}(y_j))\}, \ j = \overline{1,m}, \ p = \overline{1,q_j},$$

where $\mu^{c_{il}}(x_i)$ is a membership function of variable x_i to the term-assessment c_{il}, $i = \overline{1,n}$, $l = \overline{1,k_i}$; $\mu^{e_{jp}}(y_j)$ is a membership function of variable y_j to the term-assessment e_{jp}, $j = \overline{1,m}$, $p = \overline{1,q_j}$.

We shall redenote the set of input and output terms-assessments in the following way:

$\{C_1, C_2, ..., C_N\} = \{c_{11}, c_{12}, ..., c_{1k_1}, ..., c_{n1}, c_{n2}, ..., c_{nk_n}\}$ is the set of terms for input parameters evaluation, where $N = k_1 + k_2 + ... + k_n$;

$\{E_1, E_2, ..., E_M\} = \{\, e_{11}, e_{12}, ..., e_{1q_1}, ..., e_{m1}, e_{m2}, ..., e_{mq_m}\,\}$ is the set of terms for output parameters evaluation, where $M = q_1 + q_2 + ... + q_m$.

Set $\{C_I, I = \overline{1, N}\}$ is called fuzzy causes (diagnoses), and set $\{E_J,\ J = \overline{1, M}\}$ is called fuzzy effects (symptoms).

The diagnostic problem is set in the following way: it is necessary to restore and identify the values of the input parameters $(x_1^*, x_2^*, ..., x_n^*)$ through the values of the observed output parameters $(y_1^*, y_2^*, ..., y_m^*)$.

To interrelate the causes and effects, we will use a matrix of fuzzy relations R or a system of fuzzy IF-THEN rules.

3. Fuzzy Model of Diagnosis

3.1. Diagnostic Approximator Based on Fuzzy Relations

"Causes-effects" interconnection is given by the system of SISO fuzzy relational matrices $R_{ij} \subseteq c_{il} \times e_{jp} = [r_{il, jp}, i = \overline{1, n},\ j = \overline{1, m}, l = \overline{1, k_i},\ p = \overline{1, q_j}]$, which is equal to the MIMO fuzzy relational matrix $R \subseteq C_I \times E_J = [r_{IJ},\ I = \overline{1, N},\ J = \overline{1, M}]$. An element of this matrix is a number $r_{IJ} \in [0, 1]$, characterizing the degree to which cause C_I influences upon the rise of effect E_J.

Matrix R can be obtained on the basis of expert assessments [21].

Given the matrix R, the "causes-effects" dependency can be described with the help of the extended compositional rule of inference [2]

$$(\mu^{B1}, \mu^{B2}, ..., \mu^{Bm}) = (\mu^{A1}, \mu^{A2}, ..., \mu^{An}) * \begin{bmatrix} R_{11} & R_{12} & ... & R_{1m} \\ R_{21} & R_{22} & ... & R_{2m} \\ ... & ... & ... & ... \\ R_{n1} & R_{n2} & ... & R_{nm} \end{bmatrix}, \tag{1}$$

where

$(\mu^{A1}, \mu^{A2}, ... \mu^{An}) = ((\mu^{f11}, \mu^{f12}, ... \mu^{c_{1k_1}}), (\mu^{f21}, \mu^{f22}, ... \mu^{c_{2k_2}}), ... (\mu^{fn1}, \mu^{fn2}, ... \mu^{c_{nk_n}}))$ or $\mu^C = (\mu^{C1}, \mu^{C2}, ..., \mu^{CN})$ is the fuzzy causes vector with elements $\mu^{C_I} \in [0, 1]$, interpreted as some significance measures of C_I causes;

$(\mu^{B_1},\mu^{B_2},...\mu^{B_m})=((\mu^{e_{11}},\mu^{e_{12}},...\mu^{e_{1q_1}}),(\mu^{e_{21}},\mu^{e_{22}},...\mu^{e_{2q_2}}),...(\mu^{e_{m1}},\mu^{e_{m2}},...\mu^{e_{mq_n}}))$ or $\mu^{E} =$ $(\mu^{E_1},\mu^{E_2},...,\mu^{EM})$ is the fuzzy effects vector with elements $\mu^{E_J} \in [0, 1]$, interpreted as some significance measures of E_J effects;

* is the operation of ($\circ$, $\cap$) [2].

Finding vector μ^{C} amounts to the solution of the system of multidimensional fuzzy relational equations, which is derived from relation (1)

$$\mu^{B_1} = \mu^{A_1} \circ R_{11} \cap \mu^{A_2} \circ R_{21} \cap ... \mu^{A_n} \circ R_{n1}$$

$$\mu^{B_2} = \mu^{A_1} \circ R_{12} \cap \mu^{A_2} \circ R_{22} \cap ... \mu^{A_n} \circ R_{n2}$$

$$...$$

$$\mu^{B_m} = \mu^{A_1} \circ R_{1m} \cap \mu^{A_2} \circ R_{2m} \cap ... \mu^{A_n} \circ R_{nm} . \tag{2}$$

Since the operation $\circ$ is associated with *max-min* and the operation $\cap$ is replaced by *min* in fuzzy set theory [2], system (2) can be rearranged as:

$$\mu^{e_{jp}} = \min_{i=1,n} [\max_{l=1,k_i} (\min(\mu^{c_{il}} , r_{il,jp}))], \quad j=\overline{1,m}, \ p=\overline{1,q_j} . \tag{3}$$

In order to translate the specific values of the input and output variables into the measures of the causes and effects significances it is necessary to define a membership function of fuzzy terms C_I and E_J, $I=\overline{1,N}$, $J=\overline{1,M}$. We use a bell-shaped membership function model of variable u to arbitrary term T in the form [15]:

$$\mu^{T}(u) = \frac{1}{1+((u-\beta)/\sigma)^2}, \tag{4}$$

where β is a coordinate of function maximum, $\mu^{T}(\beta)=1$; σ is a parameter of concentration-extension (Figure 2).

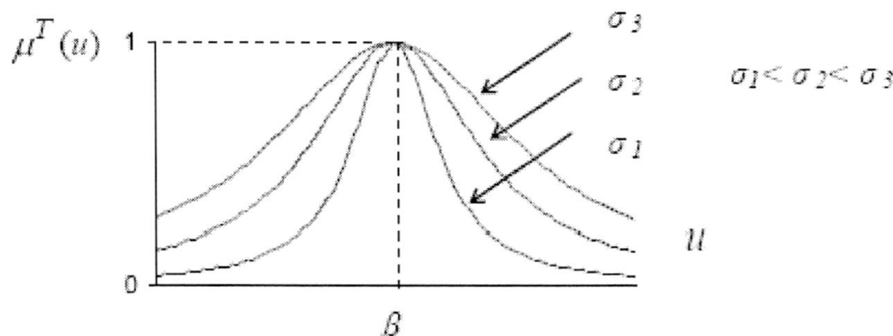

Figure 2. Model of the bell-shaped membership function.

Correlations (3) and (4) define the generalized fuzzy model of diagnosis as follows:

$$\mu^E (Y, B_E, \Omega_E) = F_R (X, R, B_C, \Omega_C), \tag{5}$$

where $B_C = (\beta^{C_1}, \beta^{C_2}, ..., \beta^{C_N})$ and $\Omega_C = (\sigma^{C_1}, \sigma^{C_2}, ..., \sigma^{C_N})$ are the vectors of β- and σ- parameters for fuzzy causes C_1, C_2,..., C_N membership functions;

$B_E = (\beta^{E_1}, \beta^{E_2}, ..., \beta^{E_M})$ and $\Omega_E = (\sigma^{E_1}, \sigma^{E_2}, ..., \sigma^{E_M})$ are the vectors of β- and σ- parameters for fuzzy effects E_1, E_2,..., E_M membership functions;

F_R is the operator of inputs-outputs connection, corresponding to formulae (3), (4).

3.2. Diagnostic Approximator Based on Fuzzy Rules

"Causes-effects" interconnection can be represented using the expert matrix of knowledge (Table 1). The fuzzy knowledge base below corresponds to this matrix:

Rule l : IF $x_1 = a_{1l}$ AND $x_2 = a_{2l}$ ... AND $x_n = a_{nl}$

THEN $y_1 = b_{1l}$ (with weight w_{1l}) AND $y_2 = b_{2l}$ (with weight w_{2l}) ...

$$\text{AND } y_m = b_{ml} \text{ (with weight } w_{ml} \text{), } l = \overline{1, K} ; \tag{6}$$

where a_{il} is a fuzzy term for variable x_i evaluation in the rule with number l ;

b_{jl} is a fuzzy term for variable y_j evaluation in the rule with number l ;

w_{jl} is a rule weight, i.e. a number in the range [0, 1], characterizing the measure of confidence of an expert relative to the statement with number jl ;

K is the number of fuzzy rules.

Table 1. Fuzzy knowledge base

Rule	Inputs				Outputs							
	x_1	x_2	...	x_n	y_1	Weight	y_2	Weight		y_m	Weight
1	a_{11}	a_{21}	...	a_{n1}	b_{11}	w_{11}	b_{21}	w_{21}	...	b_{m1}	w_{m1}
2	a_{12}	a_{22}	...	a_{n2}	b_{12}	w_{12}	b_{22}	w_{22}	...	b_{m2}	w_{m2}
...	...	...	...	...	...	...	...	...	...	...	...
K	a_{1K}	a_{2K}	...	a_{nK}	b_{1K}	w_{1K}	b_{2K}	w_{2K}	...	b_{mK}	w_{mK}

This fuzzy rule base is modelled by the fuzzy relational matrices presented in Table 2. These relational matrices can be translated as a set of fuzzy IF-THEN rules:

IF $X = A_L$ (i.e., $x_1 = C_1$ (with weight v_{1L}) AND ... $x_i = C_I$ (with weight v_{IL}) ...

AND $x_n = C_N$ (with weight v_{CL}))

THEN $y_j = E_J$ (with weight w_{LJ}).

Here A_L is the combination of input terms in rule L, $L = \overline{1, K}$.

«Causes – rules – effects» interconnection is given by the hierarchical system of relational matrices $V \subseteq C_I \times A_L = [\, v_{IL}, \; I = \overline{1, N}, \; L = \overline{1, K} \,]$ and $W \subseteq A_L \times E_J = [\, w_{LJ}$, $L = \overline{1, K}$, $J = \overline{1, M} \,]$. An element of binary matrix V is the weight of term $v_{IL} \in \{0,1\}$, where $v_{IL} = 1(0)$ if term C_I is present (absent) in the causes combination A_L. An element of fuzzy relational matrix W is the weight of rule $w_{LJ} \in [0, 1]$, characterizing the degree to which causes combination A_L influences upon the rise of effect E_J.

Given the matrices W and V, the «causes-effects» dependency can be described with the help of Zadeh's compositional rule of inference [1]

$$\mu^E = \mu^A \circ R, \tag{7}$$

where

$$\mu^A = \mu^C \bullet \overline{V}. \tag{8}$$

Here $\overline{V}$ is the complement of the matrix of terms weights V;

$\mu^C = (\mu^{C_1}, \mu^{C_2}, ..., \mu^{C_N})$ is the fuzzy causes vector with elements $\mu^{C_I} \in [0, 1]$, interpreted as some significance measures of C_I causes;

Table 2. Fuzzy relational matrices

<table>
<tr>
<td colspan="12">IF inputs</td>
<td colspan="11">THEN outputs</td>
</tr>
<tr>
<td></td>
<td colspan="3">x_1</td>
<td>...</td>
<td colspan="3">x_i</td>
<td>...</td>
<td colspan="3">x_n</td>
<td colspan="3">y_1</td>
<td>...</td>
<td colspan="3">y_j</td>
<td>...</td>
<td colspan="3">y_m</td>
</tr>
<tr>
<td></td>
<td>c_{11}</td>
<td>...</td>
<td>c_{1k_1}</td>
<td>...</td>
<td>c_{i1}</td>
<td>...</td>
<td>c_{ik_i}</td>
<td>...</td>
<td>c_{n1}</td>
<td>...</td>
<td>c_{nk_n}</td>
<td>e_{11}</td>
<td>...</td>
<td>e_{1q_1}</td>
<td>...</td>
<td>e_{j1}</td>
<td>...</td>
<td>e_{jq_j}</td>
<td>...</td>
<td>e_{m1}</td>
<td>...</td>
<td>e_{mq_m}</td>
</tr>
<tr>
<td></td>
<td>C_1</td>
<td colspan="3">...</td>
<td>C_I</td>
<td colspan="5">...</td>
<td>C_N</td>
<td>E_1</td>
<td colspan="3">...</td>
<td>E_J</td>
<td colspan="5">...</td>
<td>E_M</td>
</tr>
<tr>
<td>A_1</td>
<td>v_{11}</td>
<td colspan="3">...</td>
<td>v_{I1}</td>
<td colspan="5">...</td>
<td>v_{N1}</td>
<td>w_{11}</td>
<td colspan="3">...</td>
<td>w_{1J}</td>
<td colspan="5">...</td>
<td>w_{1M}</td>
</tr>
<tr>
<td>...</td>
<td>...</td>
<td colspan="3">...</td>
<td>...</td>
<td colspan="5">...</td>
<td>...</td>
<td>...</td>
<td colspan="3">...</td>
<td>...</td>
<td colspan="5">...</td>
<td>...</td>
</tr>
<tr>
<td>A_L</td>
<td>v_{1L}</td>
<td colspan="3">...</td>
<td>v_{IL}</td>
<td colspan="5">...</td>
<td>v_{NL}</td>
<td>w_{L1}</td>
<td colspan="3">...</td>
<td>w_{LJ}</td>
<td colspan="5">...</td>
<td>w_{LM}</td>
</tr>
<tr>
<td>...</td>
<td>...</td>
<td colspan="3">...</td>
<td>...</td>
<td colspan="5">...</td>
<td>...</td>
<td>...</td>
<td colspan="3">...</td>
<td>...</td>
<td colspan="5">...</td>
<td>...</td>
</tr>
<tr>
<td>A_K</td>
<td>v_{1K}</td>
<td colspan="3">...</td>
<td>v_{IK}</td>
<td colspan="5">...</td>
<td>v_{NK}</td>
<td>w_{K1}</td>
<td colspan="3">...</td>
<td>w_{KJ}</td>
<td colspan="5">...</td>
<td>w_{KM}</td>
</tr>
</table>

$\mu^E = (\mu^{E_1}, \mu^{E_2}, ..., \mu^{E_M})$ is the fuzzy effects vector with elements $\mu^{E_J} \in [0, 1]$, interpreted as some significance measures of E_J effects;

$\mu^A = (\mu^{A_1}, \mu^{A_2}, ..., \mu^{A_K})$ is the fuzzy causes combinations vector with elements $\mu^{A_L} \in [0, 1]$, interpreted as some significance measures of A_L causes combinations; $\bullet$ ($\circ$) is the operation of *min-max* (*max-min*) composition [3].

Finding vector μ^C amounts to the solution of the hierarchical system of simplified fuzzy relational equations with *max-min* and dual *min-max* law of composition

$$\mu^{E_1} = (\mu^{A_1} \wedge w_{11}) \vee (\mu^{A_2} \wedge w_{21}) \vee ... \vee (\mu^{A_K} \wedge w_{K1})$$

$$\mu^{E_2} = (\mu^{A_1} \wedge w_{12}) \vee (\mu^{A_2} \wedge w_{22}) \vee ... \vee (\mu^{A_K} \wedge w_{K2})$$

...

$$\mu^{E_M} = (\mu^{A_1} \wedge w_{1M}) \vee (\mu^{A_2} \wedge w_{2M}) \vee ... \vee (\mu^{A_K} \wedge w_{KM}), \tag{9}$$

where

$$\mu^{A_1} = (\mu^{C_1} \vee \bar{v}_{11}) \wedge (\mu^{C_2} \vee \bar{v}_{21}) \wedge ... \wedge (\mu^{C_N} \vee \bar{v}_{N1})$$

$$\mu^{A_2} = (\mu^{C_1} \vee \bar{v}_{12}) \wedge (\mu^{C_2} \vee \bar{v}_{22}) \wedge ... \wedge (\mu^{C_N} \vee \bar{v}_{N2})$$

...

$$\mu^{A_K} = (\mu^{C_1} \vee \bar{v}_{1K}) \wedge (\mu^{C_2} \vee \bar{v}_{2K}) \wedge ... \wedge (\mu^{C_N} \vee \bar{v}_{NK}), \tag{10}$$

which is derived from relations (7) and (8).

Since the operations $\vee$ and $\wedge$ are replaced by *max* and *min* in fuzzy set theory [1], systems (9) and (10) can be rearranged as:

$$\mu^{E_J} = \max_{L=1,K} (min(\mu^{A_L}, w_{LJ})), \quad J = \overline{1, M}, \tag{11}$$

where

$$\mu^{A_L} = \min_{I=1,N} (max(\mu^{C_I}, \bar{v}_{IL})), \quad L = \overline{1, K} \tag{12}$$

or

$$\mu^{E_J} = \max_{L=1,K}[min(\ \min_{I=1,N}(max(\mu^{C_I},\overline{v}_{IL})),w_{LJ})],\ J=\overline{1,M}\ .\tag{13}$$

To translate the specific values of the input and output variables into the measures of the causes and effects significances it is necessary to define a membership function of linguistic terms C_I and E_J, $I=\overline{1,N}$, $J=\overline{1,M}$, used in the fuzzy rules (6). We use a bell-shaped membership function model of variable u to arbitrary term T in the form (4).

Correlations (4), (13) define the generalized fuzzy model of diagnosis as follows:

$$\mu^E(Y,B_E,\Omega_E)=F_Y(X,R,B_C,\Omega_C),\tag{14}$$

where $W=(w_{11},w_{12},...,w_{1K},...,w_{m1},w_{m2},...,w_{mK})$ is the vector of rules weights;

$B_C=(\beta^{C_1},\beta^{C_2},...,\beta^{C_N})$ and $\Omega_C=(\sigma^{C_1},\sigma^{C_2},...,\sigma^{C_N})$ are the vectors of β- and σ - parameters for fuzzy causes C_1, $C_2,...,$ C_N membership functions;

$B_E=(\beta^{E_1},\beta^{E_2},...,\beta^{E_M})$ and $\Omega_E=(\sigma^{E_1},\sigma^{E_2},...,\sigma^{E_M})$ are the vectors of β- and σ - parameters for fuzzy effects E_1, $E_2,...,E_M$ membership functions;

F_Y is the operator of inputs-outputs connection, corresponding to formulae (4), (13).

4. Solving the Least Squares Equations

4.1. Genetic Optimization of the Null Diagnostic Solution

Following the approach, proposed in [16 – 18], the problem of solving fuzzy relational equations (3) or (13) is formulated as follows. Fuzzy causes vector $\mu^C=(\mu^{C_1},\mu^{C_2},...,\mu^{C_N})$ should be found which satisfies the constraints $\mu^{C_I}\in[0,1]$, $I=\overline{1,N}$, and also provides the least distance between observed and model fuzzy effects vectors:

$$F_1=[\mu^E(Y)-F_R(\mu^C(X))]^2=\min_{\mu^C}\tag{15}$$

or

$$F_2 = [\mu^E(Y) - F_Y(\mu^C(X))]^2 = \min_{\mu^C}. \tag{16}$$

Following [16 – 18], formation of diagnostic results begins with the search for the null solution $\mu_0^C = (\mu_0^{C1}, \mu_0^{C2}, ..., \mu_0^{CN})$ of optimization problem (15) or (16). The genetic algorithm is used for the null solution finding [22, 23]. The chromosome needed in the genetic algorithm for solving these optimization problems includes the real codes of parameters μ^{CI}, $I = \overline{1, N}$. The crossover operation is carried out by way of exchanging genes inside each variable μ^{CI}. The multi-crossover operation provides a more accurate adjusting direction for evolving offsprings that allows to systematically reduce the size of the search region. The non-uniform mutation the action of which depends on the age of the population provides generation of the non-dominated solutions. We used the roulette wheel selection procedure giving priority to the best solutions. The greater the fitness function of some chromosome the greater is the probability for the given chromosome to yield offsprings. The fitness function is built on the basis of criterion (15) or (16). While performing the genetic algorithm the size of the population stays constant. That is why after crossover and mutation operations it is necessary to remove the chromosomes having the worst values of the fitness function from the obtained population.

4.2. Genetic Search for the Diagnostic Solution Set

4.2.1. Diagnosis Based on Fuzzy Relations

In the general case, system (3) has a solution set $S(R, \mu^E)$, which is completely characterized by the set of upper solutions $\overline{S}^*(R, \mu^E) = \{\overline{\mu}_k^C, k = \overline{1, T}\}$ and the set of lower solutions $\underline{S}^*(R, \mu^E) = \{\underline{\mu}_l^C, l = \overline{1, H}\}$. This implies that the solution set $S(R, \mu^E)$ contains the simplified subsets $D_k(R, \mu^E)$, $k = \overline{1, T}$, each of which is determined by the unique greatest solution $\overline{\mu}_k^C \in \overline{S}^*(R, \mu^E)$ and the set of lower solutions $\underline{S}^*(R, \mu^E) = \{\underline{\mu}_l^C, l = \overline{1, H}\}$. The solution set $S(R, \mu^E)$ is obtained by extending the simplified solution subsets $D_k(R, \mu^E)$ to the case of multidimensional fuzzy relational equations:

$$S(R, \mu^E) = \bigcup_{\overline{\mu}_k^C \in \overline{S}^*} \bigcup_{\underline{\mu}_l^C \in \underline{S}^*} \left[\underline{\mu}_l^C, \overline{\mu}_k^C \right], \, l = \overline{1, H}, \, k = \overline{1, T}. \tag{17}$$

Here $\overline{\mu}_k^C = (\overline{\mu}_k^{C_1}, \overline{\mu}_k^{C_2}, ..., \overline{\mu}_k^{C_N})$ and $\underline{\mu}_l^C = (\underline{\mu}_l^{C_1}, \underline{\mu}_l^{C_2}, ..., \underline{\mu}_l^{C_N})$ are the vectors of the

upper and lower bounds of causes C_I significance measures, where the union is taken over

all $\underline{\mu}_k^C \in \overline{S}^*(R, \mu^E)$ and $\underline{\mu}_l^C \in \underline{S}^*(R, \mu^E)$.

Formation of intervals (17) is accomplished by way of solving a multiple optimization

problem (15). The lower bound ($\underline{\mu}_l^{C_I}$) for $l = 1$ is found in the range $[0, \mu_0^{C_I}]$, and for

$l > 1$ – in the range $[0, \min\limits_{p=1,k} \overline{\mu}_p^{C_I}]$, where the minimal solutions $\underline{\mu}_s^C$, $s < l$, are

excluded from the search space. The upper bound ($\overline{\mu}_k^{C_I}$) for $k = 1$ is found in the range

$[\mu_0^{C_I}, 1]$, and for $k > 1$ – in the range $[0, \max\limits_{s=1,l} \mu_s^{C_I}]$, where the maximal solutions $\overline{\mu}_p^{C_I}$,

$p < k$, are excluded from the search space.

Let $\mu^C(t) = (\mu^{C_1}(t), \mu^{C_2}(t), ..., \mu^{C_N}(t))$ be some t-th solution of optimization

problem (15), that is $F(\mu^C(t)) = F(\mu_0^C)$, since for all $\mu^C \in S(R, \mu^E)$ we have the

same value of criterion (15). While searching for upper bounds ($\overline{\mu}_k^{C_I}$) it is suggested that

$\mu^{C_I}(t) \geq \mu^{C_I}(t-1)$, and while searching for lower bounds ($\underline{\mu}_l^{C_I}$) it is suggested that

$\mu^{C_I}(t) \leq \mu^{C_I}(t-1)$. The definition of the upper (lower) bounds follows the rule: if

$\mu^C(t) \neq \mu^C(t-1)$, then $\overline{\mu}_k^{C_I}(\underline{\mu}_l^{C_I}) = \mu^{C_I}(t)$, $I = \overline{1, N}$. If $\mu^C(t) = \mu^C(t-1)$,

then the search for the interval solution $[\underline{\mu}_l^C, \overline{\mu}_k^C]$ is stopped. Formation of intervals (17)

will go on till the conditions $\overline{\mu}_k^C \neq \overline{\mu}_p^C$ and $\underline{\mu}_l^C \neq \underline{\mu}_s^C$, $p < k$, $s < l$, have been

satisfied.

4.2.2. Diagnosis Based on Fuzzy Rules

Following [3], in the general case, system (11) has a solution set $S(R, \mu^E)$, which is

completely characterized by the unique greatest solution $\overline{\mu}^A$ and the set of lower solutions

$S^*(R, \mu^E) = \{\underline{\mu}_k^A, k = \overline{1, T}\}$:

$$S(R, \mu^E) = \bigcup_{\underline{\mu}_k^A \in S^*} \left[\underline{\mu}_k^A, \overline{\mu}^A \right]. \tag{18}$$

Here $\overline{\mu}^A = (\overline{\mu}^{A_1}, \overline{\mu}^{A_2}, ..., \overline{\mu}^{A_K})$ and $\underline{\mu}_k^A = (\underline{\mu}_k^{A_1}, \underline{\mu}_k^{A_2}, ..., \underline{\mu}_k^{A_K})$ are the vectors of the upper and lower bounds of causes combinations A_L significance measures, where the union is taken over all $\underline{\mu}_k^A \in S^*(R, \mu^E)$.

Formation of intervals (18) begins with the search for the null vector of the causes combinations significances measures $\mu_0^A(\mu_0^C) = (\mu_0^{A_1}, \mu_0^{A_2}, ..., \mu_0^{A_K})$, which corresponds to the obtained null solution $\mu_0^C = (\mu_0^{C_1}, \mu_0^{C_2}, ..., \mu_0^{C_N})$. The upper bound ($\overline{\mu}^{A_L}$) is found in the range $[\mu_0^{A_L}, 1]$. The lower bound ($\underline{\mu}_k^{A_L}$) for $k = 1$ is found in the range $[0, \mu_0^{A_L}]$, and for $k > 1$ – in the range $[0, \overline{\mu}^{A_L}]$, where the minimal solutions $\underline{\mu}_p^A$, $p < k$, are excluded from the search space. Let $\mu^A(t) = (\mu^{A_1}(t), \mu^{A_2}(t), ..., \mu^{A_K}(t))$ be some t-th solution of system (11). While searching for upper bounds ($\overline{\mu}^{A_L}$) it is suggested that $\mu^{A_L}(t) \geq \mu^{A_L}(t-1)$, and while searching for lower bounds ($\underline{\mu}_k^{A_L}$) it is suggested that $\mu^{A_L}(t) \leq \mu^{A_L}(t-1)$. The definition of the upper (lower) bounds follows the rule: if $\mu^A(t) \neq \mu^A(t-1)$, then $\overline{\mu}^{A_L}$ ($\underline{\mu}_k^{A_L}$)$= \mu^{A_L}(t)$, $L = \overline{1, K}$. If $\mu^A(t) = \mu^A(t-1)$, then the search for the interval solution $[\underline{\mu}_k^A, \overline{\mu}^A]$ is stopped. Formation of intervals (18) will go on till the condition $\underline{\mu}_k^A \neq \underline{\mu}_p^A$, $p < k$, has been satisfied.

For the greatest solution $\overline{\mu}^A$, system (12) has a solution set $\overline{D}(\overline{\mu}^A)$, which is completely characterized by the unique least solution $\underline{\mu}^C$ and the set of upper solutions $\overline{D}^*(\overline{\mu}^A) = \{\overline{\mu}_l^C, l = \overline{1, H}\}$:

$$\overline{D}(\overline{\mu}^A) = \bigcup_{\overline{\mu}_l^C \in \overline{D}^*} \left[\underline{\mu}^C, \overline{\mu}_l^C \right]. \tag{19}$$

Here $\underline{\mu}^C = (\underline{\mu}^{C_1}, \underline{\mu}^{C_2}, ..., \underline{\mu}^{C_N})$ and $\overline{\mu}_l^C = (\overline{\mu}_l^{C_1}, \overline{\mu}_l^{C_2}, ..., \overline{\mu}_l^{C_N})$ are the vectors of the lower and upper bounds of causes C_I significance measures, where the union is taken over all $\overline{\mu}_l^C \in \overline{D}^*(\overline{\mu}^A)$.

For each lower solution $\underline{\mu}_k^A$, $k = \overline{1,T}$, system (12) has a solution set $\underline{D}_k(\underline{\mu}_k^A)$, which is completely characterized by the unique least solution $\underline{\mu}_k^C$ and the set of upper solutions

$$\underline{D}_k^*(\underline{\mu}_k^A) = \left\{ \overline{\mu}_{kl}^C, l = \overline{1, H_k} \right\}:$$

$$\underline{D}_k(\underline{\mu}_k^A) = \bigcup_{\overline{\mu}_{kl}^C \in \underline{D}_k^*} \left[\underline{\mu}_k^C, \overline{\mu}_{kl}^C \right]. \tag{20}$$

Here $\underline{\mu}_k^C = (\underline{\mu}_k^{C_1}, \underline{\mu}_k^{C_2}, ..., \underline{\mu}_k^{C_N})$ and $\overline{\mu}_{kl}^C = (\overline{\mu}_{kl}^{C_1}, \overline{\mu}_{kl}^{C_2}, ..., \overline{\mu}_{kl}^{C_N})$ are the vectors of the lower and upper bounds of causes C_I significance measures, where the union is taken over all $\overline{\mu}_{kl}^C \in \underline{D}_k^*(\underline{\mu}_k^A)$.

Formation of intervals (19) begins with the search for the null solution $\overline{\mu}_0^C = (\overline{\mu}_0^{C_1}, \overline{\mu}_0^{C_2}, ..., \overline{\mu}_0^{C_N})$ for the greatest solution $\overline{\mu}^A$. The lower bound ($\underline{\mu}^{C_I}$) of solution set (19) is found in the range $[0, \overline{\mu}_0^{C_I}]$. The upper bound ($\overline{\mu}_l^{C_I}$) for $l = 1$ is found in the range $[\overline{\mu}_0^{C_I}, 1]$, and for $l > 1$ – in the range $[\underline{\mu}^{C_I}, 1]$, where the maximal solutions $\overline{\mu}_p^{C_I}$, $p < l$, are excluded from the search space.

Formation of intervals (20) begins with the search for the null solutions $\underline{\mu}_{0k}^C = (\underline{\mu}_{0k}^{C_1}, \underline{\mu}_{0k}^{C_2}, ..., \underline{\mu}_{0k}^{C_N})$ for each of the lower solutions $\underline{\mu}_k^A$, $k = \overline{1,T}$. The lower bound ($\underline{\mu}_k^{C_I}$) of solution set (20) is found in the range $[0, \underline{\mu}_{0k}^{C_I}]$. The upper bound ($\overline{\mu}_{kl}^{C_I}$) for $l = 1$ is found in the range $[\underline{\mu}_{0k}^{C_I}, 1]$, and for $l > 1$ – in the range $[\underline{\mu}_k^{C_I}, 1]$, where the maximal solutions $\overline{\mu}_{kp}^{C_I}$, $p < l$, are excluded from the search space.

Let $\mu^C(t) = (\mu^{C_1}(t), \mu^{C_2}(t), ..., \mu^{C_N}(t))$ be some t-th solution of system (12). While searching for upper bounds ($\overline{\mu}_l^{C_I}$ or $\overline{\mu}_{kl}^{C_I}$) it is suggested that

$\mu^{CI}(t) \geq \mu^{CI}(t-1)$, and while searching for lower bounds ($\underline{\mu}^{CI}$ or $\underline{\mu}_k^{CI}$) it is

suggested that $\mu^{CI}(t) \leq \mu^{CI}(t-1)$. The definition of the upper (lower) bounds follows the

rule: if $\mu^{C}(t) \neq \mu^{C}(t-1)$, then $\overline{\mu}_l^{CI}(\underline{\mu}^{CI}) = \mu^{CI}(t)$ or $\overline{\mu}_{kl}^{CI}(\underline{\mu}_k^{CI}) = \mu^{CI}(t)$,

$I = \overline{1,N}$. If $\mu^{C}(t) = \mu^{C}(t-1)$, then the search for the interval solution $[\underline{\mu}_l^{C}, \overline{\mu}^{C}]$ or

$[\underline{\mu}_{kl}^{C}, \overline{\mu}_k^{C}]$ is stopped. Formation of intervals (19) and (20) will go on till the conditions

$\overline{\mu}_l^{C} \neq \overline{\mu}_p^{C}$ and $\overline{\mu}_{kl}^{C} \neq \overline{\mu}_{kp}^{C}$, $p < l$, have been satisfied.

5. Fuzzy Model Tuning

It is assumed that the training data which is given in the form of L pairs of experimental

data is known: $\left\langle \hat{X}_p, \hat{Y}_p \right\rangle$, $p = \overline{1,L}$, where $\hat{X}_p = (\hat{x}_1^p, \hat{x}_2^p, ..., \hat{x}_n^p)$ and

$\hat{Y}_p = (\hat{y}_1^p, \hat{y}_2^p, ..., \hat{y}_m^p)$ are the vectors of the values of the input and output variables in the
experiment number p.

The essence of tuning of the fuzzy models (5) and (14) consists of finding such matrix R
(W) and such vectors of membership functions parameters B_C, Ω_C, B_E, Ω_E, which
provide the least distance between model and experimental fuzzy effects vectors:

$$\sum_{p=1}^{L} [F_R(\hat{X}_p, R, B_C, \Omega_C) - \hat{\mu}^E(\hat{Y}_p, B_E, \Omega_E)]^2 = \min_{R, B_C, \Omega_C, B_E, \Omega_E}$$

$$(21)$$

or

$$\sum_{p=1}^{L} [F_Y(\hat{X}_p, W, B_C, \Omega_C) - \hat{\mu}^E(\hat{Y}_p, B_E, \Omega_E)]^2 = \min_{W, B_C, \Omega_C, B_E, \Omega_E}$$

$$(22)$$

The chromosome needed in the multi-crossover real-coded genetic algorithm [22, 23] for
solving these optimization problems is defined as the vector of real parameters R (W), B_C,
Ω_C, B_E, Ω_E. Fitness function is built on the basis of criterion (21) or (22).

6. Computer Simulations

The aim of the experiment consists of checking the performance of the above proposed models and algorithms of diagnosis with the help of the target "two inputs (x_1, x_2) – two outputs (y_1, y_2)" model. The target models were some analytical functions $y_1 = f_1(x_1, x_2)$ and $y_2 = f_2(x_1, x_2)$, which were approximated by the rule of inference (1) and the system of fuzzy rules (6). These functions served simultaneously as training and testing data generator. The input values (x_1, x_2) restored for each output combination (y_1, y_2) were compared with the target level lines.

The target models are given by the formulae: $y_1 = ((2z - 0.9)(7z - 1)(17z - 19)(15z - 2))/10$, $y_2 = -y_1/2 + 1$, where $z = ((x_1 - 3.0)^2 + (x_2 - 2.5)^2)/40$. The target model is represented in Figure 3.

The total number of the input and output terms-assessments consists of: c_{11} *Low (L)*, c_{12} *Average (A)*, c_{13} *High (H)* for x_1, c_{21} *(Low)*, c_{22} *(High)* for x_2; e_{11} *higher than Low (hL)*, e_{12} *higher than Average (hA)*, e_{13} *High (H)* for y_1; e_{21} *Low (L)*, e_{22} *lower than Average (lA)*, e_{23} *Average (A)* for y_2.

We shall define the set of causes and effects in the following way: $\{C_1, C_2, ..., C_5\} = \{c_{11}, c_{12}, c_{13}, c_{21}, c_{22}\}$; $\{E_1, E_2, ..., E_6\} = \{e_{11}, e_{12}, e_{13}, e_{21}, e_{22}, e_{23}\}$.

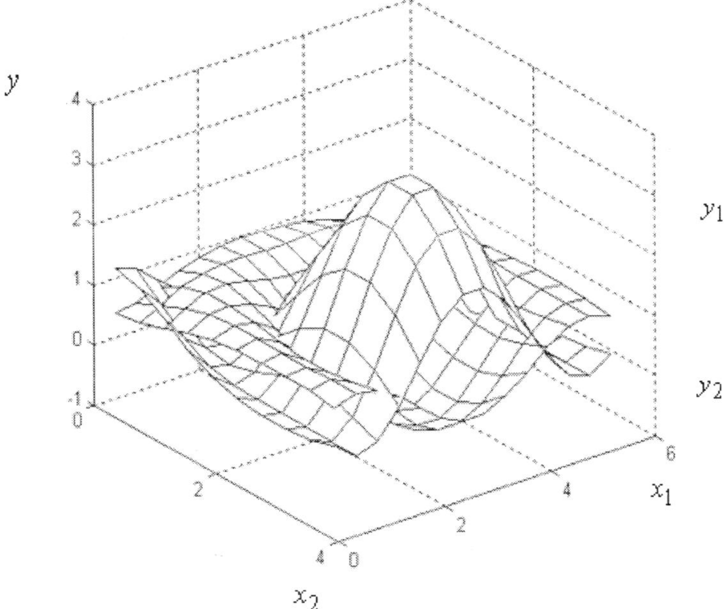

Figure 3. "Inputs-outputs" model-generator.

Table 3. Parameters of the membership functions for the causes fuzzy terms before (after) tuning

Parameter	Fuzzy terms				
	C_1	C_2	C_3	C_4	C_5
β -	0 (0.01)	3.0 (3.00)	6.0 (5.99)	0 (0.02)	3.0 (3.05)
σ -	1.0 (0.69)	2.0 (0.92)	1.0 (0.70)	1.0 (0.63)	2.0 (0.81)

Table 4. Parameters of the membership functions for the effects fuzzy terms before (after) tuning

Parameter	Fuzzy terms					
	E_1	E_2	E_3	E_4	E_5	E_6
β -	0 (0.02)	1.0 (1.12)	3.5 (3.29)	-0.7 (-0.65)	0.5 (0.41)	0.8 (0.86)
σ -	0.5 (0.26)	0.5 (0.33)	2.0 (1.88)	2.0 (1.52)	0.5 (0.37)	0.5 (0.21)

6.1. Approximation Based on Fuzzy Relations

The target model is described with the help of the following fuzzy relational model

$$(\mu^{A_1},\mu^{A_2}) * \begin{bmatrix} R_{11} & R_{12} \\ R_{21} & R_{22} \end{bmatrix} = (\mu^{B_1},\mu^{B_2}), \tag{23}$$

where $\mu^{A_1} = (\mu^{c_{11}},\mu^{c_{12}},\mu^{c_{13}})$, $\mu^{A_2} = (\mu^{c_{21}},\mu^{c_{22}})$, $\mu^{B_1} = (\mu^{e_{11}},\mu^{e_{12}},\mu^{e_{13}})$, $\mu^{B_2} = (\mu^{e_{21}},\mu^{e_{22}},\mu^{e_{23}})$.

The resulting expert fuzzy relational matrix takes the following form:

$$
R = \begin{array}{c} \\ c_{11} \\ c_{12} \\ c_{13} \\ \\ c_{21} \\ c_{22} \end{array}
\begin{array}{c} e_{11}\ e_{12}\ e_{13}\quad e_{21}\ e_{22}\ e_{23} \\ \left[\begin{array}{cc} R_{11} & R_{12} \\ & \\ R_{21} & R_{22} \end{array}\right] \end{array}
=
\begin{array}{c} \\ C_1 \\ C_2 \\ C_3 \\ C_4 \\ C_5 \end{array}
\begin{array}{c} E_1\quad E_2\quad E_3\qquad E_4\quad E_5\quad E_6 \\ \left[\begin{array}{cccccc} 0.11 & 0.89 & 0.11 & 0.55 & 0.89 & 0.11 \\ 0.99 & 0.44 & 0.99 & 0.78 & 0.22 & 0.89 \\ 0.11 & 0.89 & 0.11 & 0.55 & 0.89 & 0.11 \\ 0.99 & 0.89 & 0.11 & 0.22 & 0.89 & 0.89 \\ 0.22 & 0.33 & 0.99 & 0.78 & 0.44 & 0.33 \end{array}\right] \end{array}.
$$

The results of the fuzzy model tuning are given in Tables 3 and 4.

The system of multidimensional fuzzy relational equations is derived from relation (23). Diagnostic equations after tuning take the following form:

$$\mu^{E_1} = [(\mu^{C_1} \wedge 0.12) \vee (\mu^{C_2} \wedge 0.99) \vee (\mu^{C_3} \wedge 0.12)] \wedge [(\mu^{C_4} \wedge 0.99) \vee (\mu^{C_5} \wedge 0.24)]$$

$$\mu^{E_2} = [(\mu^{C_1} \wedge 0.88) \vee (\mu^{C_2} \wedge 0.41) \vee (\mu^{C_3} \wedge 0.88)] \wedge [(\mu^{C_4} \wedge 0.88) \vee (\mu^{C_5} \wedge 0.30)]$$

$$\mu^{E_3} = [(\mu^{C_1} \wedge 0.11) \vee (\mu^{C_2} \wedge 0.99) \vee (\mu^{C_3} \wedge 0.11)] \wedge [(\mu^{C_4} \wedge 0.11) \vee (\mu^{C_5} \wedge 0.99)]$$

$$\mu^{E_4} = [(\mu^{C_1} \wedge 0.50) \vee (\mu^{C_2} \wedge 0.75) \vee (\mu^{C_3} \wedge 0.50)] \wedge [(\mu^{C_4} \wedge 0.18) \vee (\mu^{C_5} \wedge 0.75)]$$

$$\mu^{E_5} = [(\mu^{C_1} \wedge 0.89) \vee (\mu^{C_2} \wedge 0.21) \vee (\mu^{C_3} \wedge 0.89)] \wedge [(\mu^{C_4} \wedge 0.89) \vee (\mu^{C_5} \wedge 0.43)]$$

$$\mu^{E_6} = [(\mu^{C_1} \wedge 0.13) \vee (\mu^{C_2} \wedge 0.90) \vee (\mu^{C_3} \wedge 0.13)] \wedge [(\mu^{C_4} \wedge 0.90) \vee (\mu^{C_5} \wedge 0.36)]$$

$$(24)$$

The results of solving the problem of inverse inference before and after tuning are shown in Figure 4. The same figure depicts the membership functions of the fuzzy terms for the causes and effects before and after tuning.

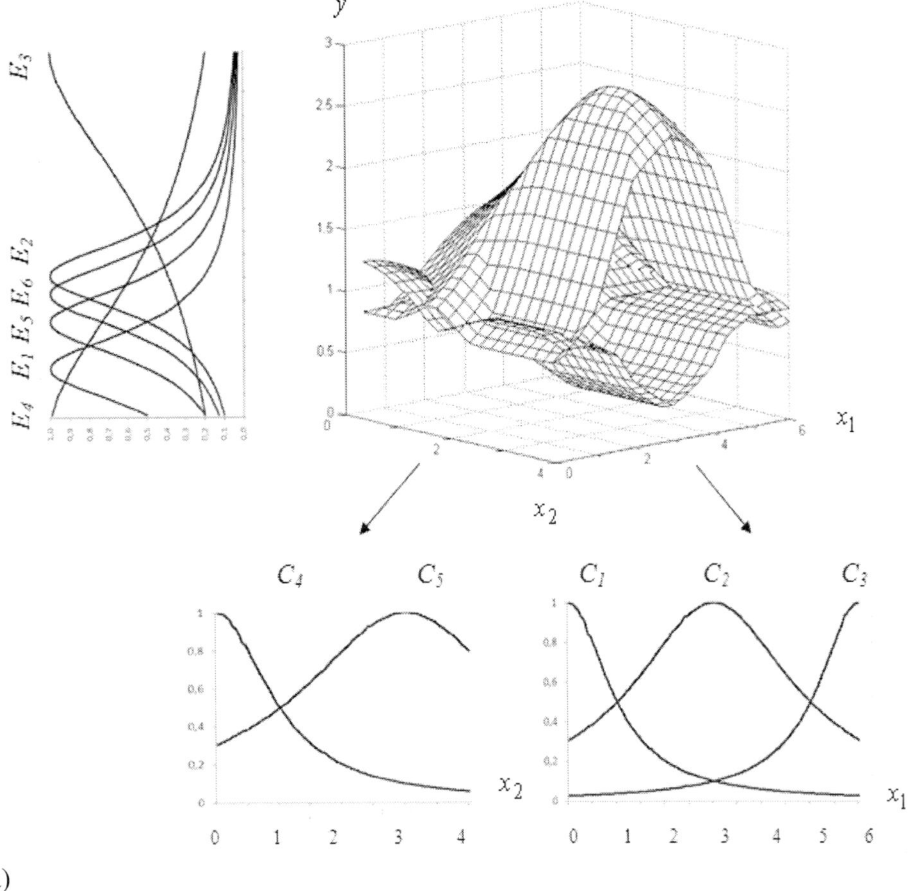

Figure 4. (Continued).

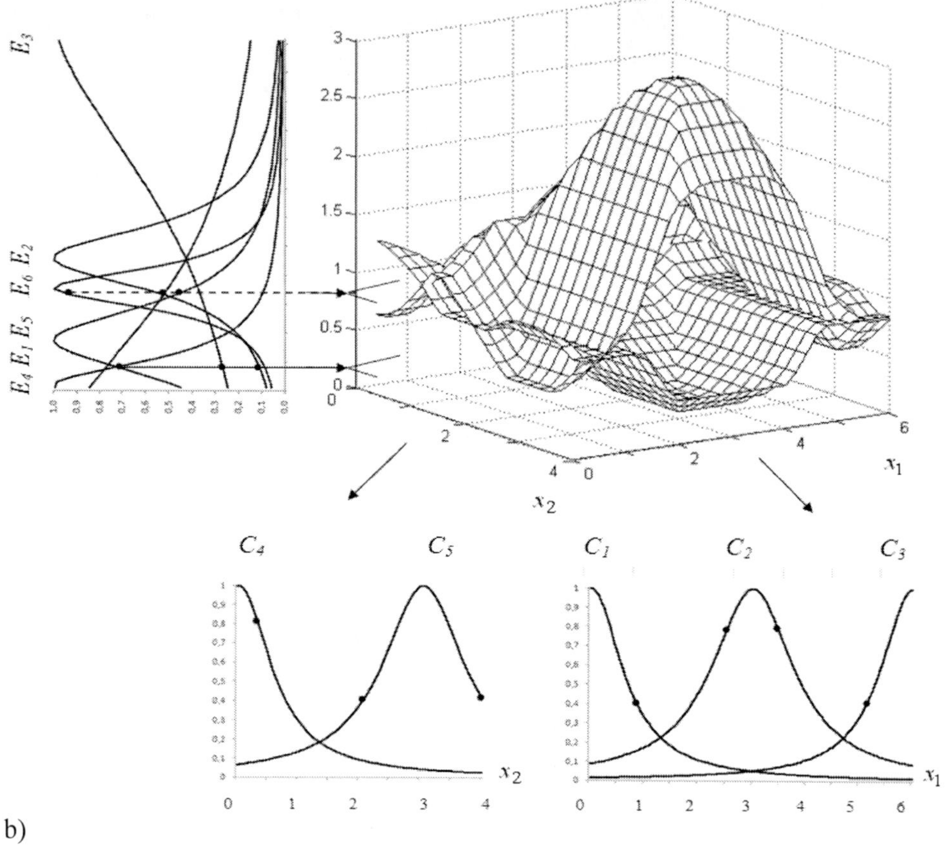

Figure 4. Solution to the problem of inverse fuzzy inference: (a) before tuning; (b) after tuning.

Let the specific values of the output variables consists of $y_1^* =0.20$ and $y_2^* =0.80$. The measures of the effects significances for these values can be defined with the help of the membership functions in Figure 4,b:

$$\mu^E =(\ \mu^{E_1}(y_1^*) =0.68;\ \mu^{E_2}(y_1^*) =0.11;\ \mu^{E_3}(y_1^*) =0.27;$$

$$\mu^{E_4}(y_2^*) =0.52;\ \mu^{E_5}(y_2^*) =0.47;\ \mu^{E_6}(y_2^*) =0.92).$$

The genetic algorithm yields a null solution of the system (24)

$$\mu_0^C =(\mu_0^{C_1} =0.41,\ \mu_0^{C_2} =0.80,\ \mu_0^{C_3} =0.26,\ \mu_0^{C_4} =0.92,\ \mu_0^{C_5} =0.39),$$

for which the optimization criterion (15) takes the value of $F_1 =0.1512$.

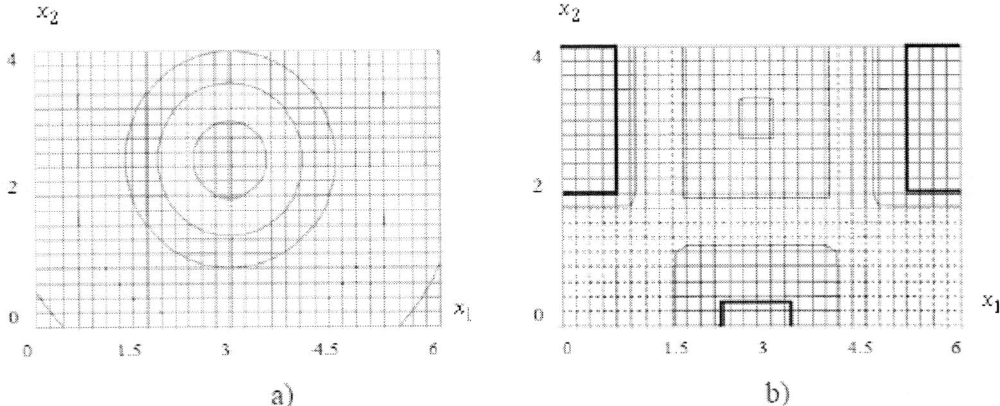

Figure 5. Comparison of the target (a) and restored (b) level lines for $y_1^* = 0.20$ and $y_2^* = 0.80$.

The obtained null solution allows us to arrange for the genetic search for the solution set $S(R, \mu^E)$, which is completely determined by the two upper solutions $\overline{S}^* = \{\overline{\mu}_1^C, \overline{\mu}_2^C\}$

$$\overline{\mu}_1^C = (\ \overline{\mu}_1^{C_1} = 0.41,\ \overline{\mu}_1^{C_2} = 0.80,\ \overline{\mu}_1^{C_3} = 0.41,\ \overline{\mu}_1^{C_4} = 1.0,\ \overline{\mu}_1^{C_5} = 0.39);$$

$$\overline{\mu}_2^C = (\ \overline{\mu}_2^{C_1} = 0.41,\ \overline{\mu}_2^{C_2} = 1.0,\ \overline{\mu}_2^{C_3} = 0.41,\ \overline{\mu}_2^{C_4} = 0.8,\ \overline{\mu}_2^{C_5} = 0.39),$$

and the two lower solutions $\underline{S}^* = \{\underline{\mu}_1^C, \underline{\mu}_2^C\}$

$$\underline{\mu}_1^C = (\ \underline{\mu}_1^{C_1} = 0.41,\ \underline{\mu}_1^{C_2} = 0.80,\ \underline{\mu}_1^{C_3} = 0,\ \underline{\mu}_1^{C_4} = 0.80,\ \underline{\mu}_1^{C_5} = 0.39);$$

$$\underline{\mu}_2^C = (\ \underline{\mu}_2^{C_1} = 0,\ \underline{\mu}_2^{C_2} = 0.80,\ \underline{\mu}_2^{C_3} = 0.41,\ \underline{\mu}_2^{C_4} = 0.80,\ \underline{\mu}_2^{C_5} = 0.39).$$

Thus, the solution set of fuzzy relational equations (24) contains the two subsets $D_1(R, \mu^E)$ and $D_2(R, \mu^E)$, each of which can be represented in the form of intervals:

$$D_1 = \{\ \mu^{C_1} = 0.41;\ \mu^{C_2} = 0.80;\ \mu^{C_3} \in [0, 0.41];\ \mu^{C_4} \in [0.80, 1.0];\ \mu^{C_5} = 0.39\}$$

$$\cup\ \{\ \mu^{C_1} \in [0, 0.41];\ \mu^{C_2} = 0.80;\ \mu^{C_3} = 0.41;\ \mu^{C_4} \in [0.80, 1.0];\ \mu^{C_5} = 0.39\},$$

$$\tag{25}$$

$$D_2 = \{\ \mu^{C_1} = 0.41;\ \mu^{C_2} \in [0.80, 1.0];\ \mu^{C_3} [0, 0.41];\ \mu^{C_4} = 0.80;\ \mu^{C_5} = 0.39\}$$

$$\cup \{ \mu^{C_1} \in [0, 0.41]; \ \mu^{C_2} \in [0.80, 1.0]; \ \mu^{C_3} = 0.41; \ \mu^{C_4} = 0.80; \ \mu^{C_5} = 0.39 \}.$$

$$(26)$$

The intervals of the values of the input variables for each interval in solutions (25) and (26) can be defined with the help of the membership functions in Figure 4,b:

- $x_1^* = 0.85$ or $x_1^* \in [0.85, 6.0]$ for C_1 ;

- $x_1^* = 2.55$, $x_1^* = 3.45$ or $x_1^* \in [2.55, 3.45]$ for C_2 ;

- $x_1^* = 5.15$ or $x_1^* \in [0, 5.15]$ for C_3 ;

- $x_2^* \in [0, 0.33]$ or $x_2^* = 0.33$ for C_4 ;

- $x_2^* = 2.04$ or $x_2^* = 4.05$ for C_5 .

The restoration of the input set for $y_1^* = 0.20$ and $y_2^* = 0.80$ is shown in Figure 4,b, in which the values of the causes and effects significances measures are marked. The comparison of the target and restored level lines for $y_1^* = 0.20$ and $y_2^* = 0.80$ is shown in Figure 5.

6.2. Approximation Based on Fuzzy Rules

The fuzzy IF-THEN rules correspond to the target model:

Rule 1: IF $x_1 = c_{11}\,(L)$ AND $x_2 = c_{21}\,(L)$ THEN $y_1 = e_{12}\,(hA)$ AND $y_2 = e_{22}\,(lA)$;

Rule 2: IF $x_1 = c_{12}\,(A)$ AND $x_2 = c_{21}\,(L)$ THEN $y_1 = e_{11}\,(hL)$ AND $y_2 = e_{23}\,(A)$;

Rule 3: IF $x_1 = c_{13}\,(H)$ AND $x_2 = c_{21}\,(L)$ THEN $y_1 = e_{12}\,(hA)$ AND $y_2 = e_{22}\,(lA)$;

Rule 4: IF $x_1 = c_{11}\,(L)$ AND $x_2 = c_{22}\,(H)$ THEN $y_1 = e_{11}\,(hL)$ AND $y_2 = e_{23}\,(A)$;

Rule 5: IF $x_1 = c_{12}\,(A)$ AND $x_2 = c_{22}\,(H)$ THEN $y_1 = e_{13}\,(H)$ AND $y_2 = e_{21}\,(L)$;

Rule 6: IF $x_1 = c_{13}\,(H)$ AND $x_2 = c_{22}\,(H)$ THEN $y_1 = e_{11}\,(hL)$ AND $y_2 = e_{23}\,(A)$.

Table 5. Fuzzy knowledge matrix

IF inputs			THEN outputs					
	x_1	x_2	y_1			y_2		
			hL	hA	H	L	lA	A
A_1	L	L	0	1	0	0	1	0
A_2	A	L	1	0	0	0	0	1
A_3	H	L	0	1	0	0	1	0
A_4	L	H	1	0	0	0	0	1
A_5	A	H	0	0	1	1	0	0
A_6	H	H	1	0	0	0	0	1

**Table 6. Parameters of the membership functions for the causes
fuzzy terms before (after) tuning**

Parameter	Fuzzy terms				
	C_1	C_2	C_3	C_4	C_5
β -	0 (0.03)	3.0 (3.03)	6.0 (5.98)	0 (0.02)	3.0 (3.05)
σ -	1.0 (0.71)	2.0 (0.62)	1.0 (0.69)	1.0 (0.73)	2.0 (0.60)

**Table 7. Parameters of the membership functions for the effects
fuzzy terms before (after) tuning**

Parameter	Fuzzy terms					
	E_1	E_2	E_3	E_4	E_5	E_6
β -	0 (0.02)	1.0 (1.10)	3.5 (3.36)	-0.7 (-0.67)	0.5 (0.44)	0.8 (0.89)
σ -	0.5 (0.27)	0.5 (0.29)	2.0 (1.91)	2.0 (1.70)	0.5 (0.31)	0.5 (0.25)

This fuzzy rule base is modelled by the fuzzy relational matrix presented in Table 5.
The results of the fuzzy model tuning are given in Tables 6 and 7.
Fuzzy relational equations after tuning take the following form:

$$\mu^{E1} = (\mu^{A2} \wedge 0.75) \vee (\mu^{A4} \wedge 0.78) \vee (\mu^{A6} \wedge 0.86)$$

$$\mu^{E2} = (\mu^{A1} \wedge 0.80) \vee (\mu^{A3} \wedge 0.92)$$

$$\mu^{E3} = (\mu^{A5} \wedge 0.97)$$

$$\mu^{E4} = (\mu^{A1} \wedge 0.50) \vee (\mu^{A3} \wedge 0.48) \vee (\mu^{A5} \wedge 0.77)$$

$$\mu^{E5} = (\mu^{A1} \wedge 0.76) \vee (\mu^{A3} \wedge 0.72)$$

$$\mu^{E6} = (\mu^{A2} \wedge 0.96) \vee (\mu^{A4} \wedge 0.82) \vee (\mu^{A6} \wedge 0.87) \tag{27}$$

where

$$\mu^{C1} \wedge \mu^{C4} = \mu^{A1}$$

$$\mu^{C2} \wedge \mu^{C4} = \mu^{A2}$$

$$\mu^{C3} \wedge \mu^{C4} = \mu^{A3}$$

$$\mu^{C1} \wedge \mu^{C5} = \mu^{A4}$$

$$\mu^{C2} \wedge \mu^{C5} = \mu^{A5}$$

$$\mu^{C3} \wedge \mu^{C5} = \mu^{A6}. \tag{28}$$

The results of solving the problem of inverse inference before and after tuning are shown in Figure 6. The same figure depicts the membership functions of the fuzzy terms for the causes and effects before and after tuning.

Let the specific values of the output variables consist of $y_1^* = 0.20$ and $y_2^* = 0.80$. The fuzzy effects vector for these values can be defined with the help of the membership functions in Figure 6,b:

$$\mu^E = (\mu^{E1}(y_1^*) = 0.69; \ \mu^{E2}(y_1^*) = 0.09; \ \mu^{E3}(y_1^*) = 0.27;$$

$$\mu^{E4}(y_2^*) = 0.57; \ \mu^{E5}(y_2^*) = 0.43; \ \mu^{E6}(y_2^*) = 0.89).$$

The genetic algorithm yields a null solution of the system (27)

$$\mu_0^C = (\mu_0^{C1} = 0.26, \ \mu_0^{C2} = 0.93, \ \mu_0^{C3} = 0.20, \ \mu_0^{C4} = 0.89, \ \mu_0^{C5} = 0.42),$$

for which the optimization criterion (16) takes the value of $F_2 = 0.1064$.

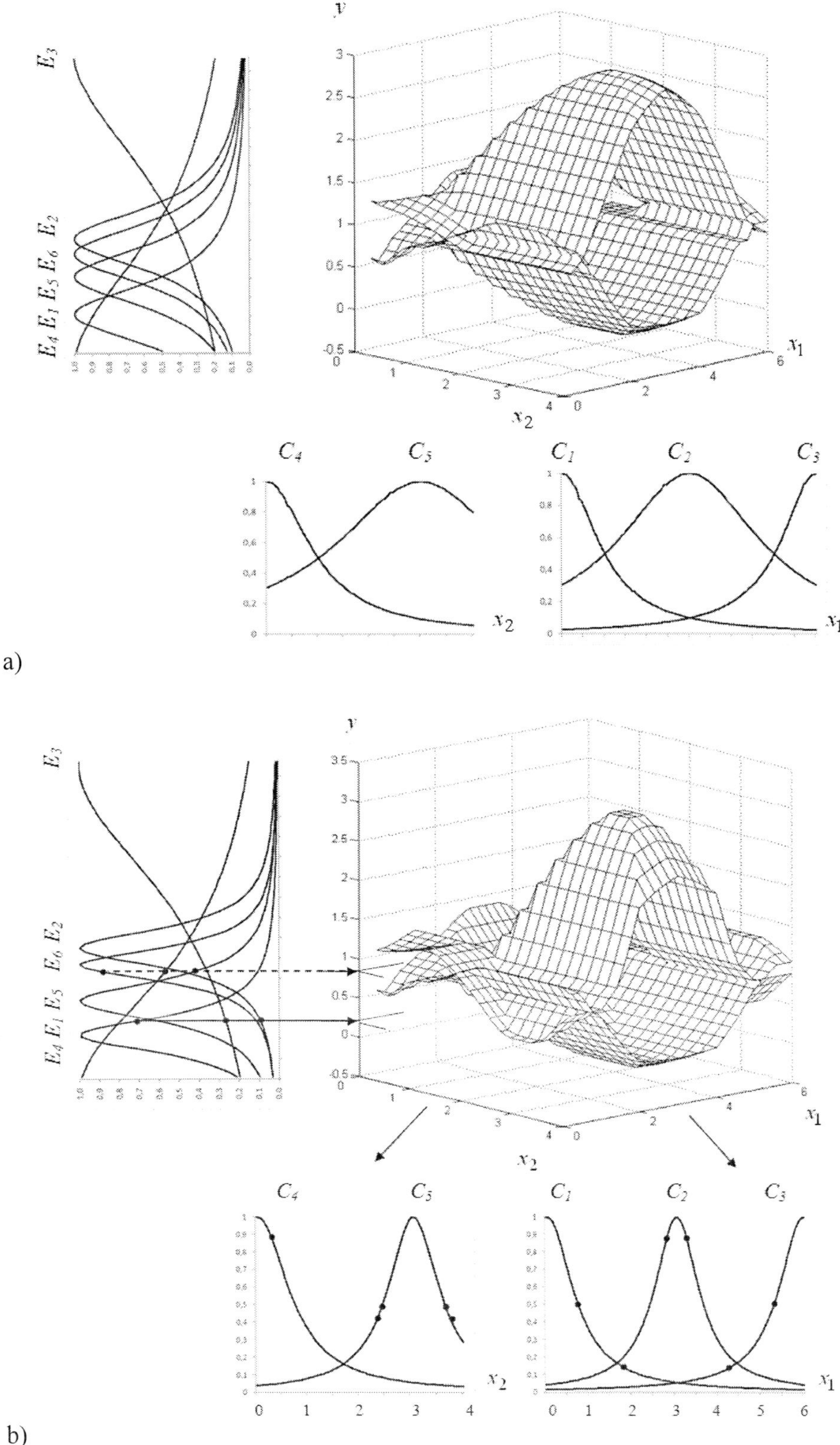

Figure 6. Solution to the problem of inverse fuzzy inference: (a) before tuning; (b) after tuning.

The null vector of the causes combinations significances measures

$$\mu_0^A = (\mu_0^{A_1} = 0.26,\ \mu_0^{A_2} = 0.89,\ \mu_0^{A_3} = 0.20,\ \mu_0^{A_4} = 0.26,\ \mu_0^{A_5} = 0.42,\ \mu_0^{A_6} = 0.20)$$

corresponds to the obtained null solution.

The obtained null solution allows us to arrange for the genetic search for the solution set $S(R, \mu^E)$, which is completely determined by the greatest solution

$$\overline{\mu}^A = (\ \overline{\mu}^{A_1} = 0.26,\ \overline{\mu}^{A_2} = 0.89,\ \overline{\mu}^{A_3} = 0.26,\ \overline{\mu}^{A_4} = 0.75,\ \overline{\mu}^{A_5} = 0.42,\ \overline{\mu}^{A_6} = 0.75)$$

and the two lower solutions $S^* = \{\underline{\mu}_1^A, \underline{\mu}_2^A\}$

$$\underline{\mu}_1^A = (\underline{\mu}_1^{A_1} = 0.26,\ \underline{\mu}_1^{A_2} = 0.89,\ \underline{\mu}_1^{A_3} = 0,\ \underline{\mu}_1^{A_4} = 0,\ \underline{\mu}_1^{A_5} = 0.42,\ \underline{\mu}_1^{A_6} = 0);$$

$$\underline{\mu}_2^A = (\underline{\mu}_2^{A_1} = 0,\ \underline{\mu}_2^{A_2} = 0.89,\ \underline{\mu}_2^{A_3} = 0.26,\ \underline{\mu}_2^{A_4} = 0,\ \underline{\mu}_2^{A_5} = 0.42,\ \underline{\mu}_2^{A_6} = 0).$$

Thus, the solution of fuzzy relational equations (27) can be represented in the form of intervals:

$$S(R, \mu^E) = \{\ \mu^{A_1} = 0.26,\ \mu^{A_2} = 0.89,\ \mu^{A_3} \in [0,\ 0.26],\ \mu^{A_4} \in [0,\ 0.75],\ \mu^{A_5} = 0.42,$$

$$\mu^{A_6} \in [0,\ 0.75]\}$$

$$\bigcup \{\ \mu^{A_1} \in [0,\ 0.26],\ \mu^{A_2} = 0.89,\ \mu^{A_3} = 0.26,\ \mu^{A_4} \in [0,\ 0.75],\ \mu^{A_5} = 0.42,\ \mu^{A_6} \in [0,\ 0.75]\}.$$

We next apply the genetic algorithm for solving (28) for the greatest solution $\overline{\mu}^A$ and the two lower solutions $\underline{\mu}_1^A$ and $\underline{\mu}_2^A$.

For the greatest solution $\overline{\mu}^A$, the genetic algorithm yields a null solution of the system (28)

$$\overline{\mu}_0^C = (\overline{\mu}_0^{C_1} = 0.49,\ \overline{\mu}_0^{C_2} = 0.96,\ \overline{\mu}_0^{C_3} = 0.49,\ \overline{\mu}_0^{C_4} = 0.90,\ \overline{\mu}_0^{C_5} = 0.49),$$

for which the optimization criterion (16) takes the value of $F_2 = 0.2221$.

The obtained null solution allows us to arrange for the genetic search for the solution set $\overline{D}(\overline{\mu}^A)$, which is completely determined by the least solution

$$\underline{\mu}^C =(\ \underline{\mu}^{C_1} =0.49,\ \underline{\mu}^{C_2} =0.89,\ \underline{\mu}^{C_3} =0.49,\ \underline{\mu}^{C_4} =0.89,\ \underline{\mu}^{C_5} =0.49)$$

and the two upper solutions $\overline{D}^* = \{\overline{\mu}_1^C, \overline{\mu}_2^C\}$

$$\overline{\mu}_1^C =(\ \overline{\mu}_1^{C_1} =0.49,\ \overline{\mu}_1^{C_2} =0.89,\ \overline{\mu}_1^{C_3} =0.49,\ \overline{\mu}_1^{C_4} =1.0,\ \overline{\mu}_1^{C_5} =0.49);$$

$$\overline{\mu}_2^C =(\ \overline{\mu}_2^{C_1} =0.49,\ \overline{\mu}_2^{C_2} =1.0,\ \overline{\mu}_2^{C_3} =0.49,\ \overline{\mu}_2^{C_4} =0.89,\ \overline{\mu}_2^{C_5} =0.49).$$

Thus, the solution of fuzzy relational equations (28) for the greatest solution $\overline{\mu}^A$ can be represented in the form of intervals:

$$\overline{D}(\overline{\mu}^A)=\{\ \mu^{C_1} =0.49,\ \mu^{C_2} =0.89,\ \mu^{C_3} =0.49,\ \mu^{C_4} \in [0.89,\ 1.0],\ \mu^{C_5} =0.49\}$$

$$\bigcup \{\ \mu^{C_1} =0.49,\ \mu^{C_2} \in [0.89,\ 1.0],\ \mu^{C_3} =0.49,\ \mu^{C_4} =0.89,\ \mu^{C_5} =0.49\}.$$

$$(29)$$

For the first lower solution $\underline{\mu}_1^A$, the genetic algorithm yields a null solution of the system (28)

$$\underline{\mu}_{01}^C =(\underline{\mu}_{01}^{C_1} = 0.13,\ \underline{\mu}_{01}^{C_2} = 0.89,\ \underline{\mu}_{01}^{C_3} = 0,\ \underline{\mu}_{01}^{C_4} = 0.94,\ \underline{\mu}_{01}^{C_5} = 0.42),$$

for which the optimization criterion (16) takes the value of $F_2 =0.1402$.

The obtained null solution allows us to arrange for the genetic search for the solution set $\underline{D}_1(\underline{\mu}_1^A)$, which is completely determined by the least solution

$$\underline{\mu}^C =(\ \underline{\mu}^{C_1} =0.13,\ \underline{\mu}^{C_2} =0.89,\ \underline{\mu}^{C_3} =0,\ \underline{\mu}^{C_4} =0.89,\ \underline{\mu}^{C_5} =0.42)$$

and the two upper solutions $\underline{D}_1^* = \{\overline{\mu}_1^C, \overline{\mu}_2^C\}$

$$\overline{\mu}_1^C =(\ \overline{\mu}_1^{C_1} =0.13,\ \overline{\mu}_1^{C_2} =0.89,\ \overline{\mu}_1^{C_3} =0,\ \overline{\mu}_1^{C_4} =1.0,\ \overline{\mu}_1^{C_5} =0.42);$$

$$\overline{\mu}_2^{C} = (\overline{\mu}_2^{C_1} = 0.13, \ \overline{\mu}_2^{C_2} = 1.0, \ \overline{\mu}_2^{C_3} = 0, \ \overline{\mu}_2^{C_4} = 0.89, \ \overline{\mu}_2^{C_5} = 0.42).$$

Thus, the solution of fuzzy relational equations (28) for the first lower solution $\underline{\mu}_1^{A}$ can be represented in the form of intervals:

$$\underline{D}_1(\underline{\mu}_1^{A}) = \{ \mu^{C_1} = 0.13, \ \mu^{C_2} = 0.89, \ \mu^{C_3} = 0, \ \mu^{C_4} \in [0.89, \ 1.0], \ \mu^{C_5} = 0.42 \}$$

$$\bigcup \{ \mu^{C_1} = 0.13, \ \mu^{C_2} \in [0.89, \ 1.0], \ \mu^{C_3} = 0, \ \mu^{C_4} = 0.89, \ \mu^{C_5} = 0.42 \}. \ (30)$$

For the second lower solution $\underline{\mu}_2^{A}$, the genetic algorithm yields a null solution of the system (28)

$$\underline{\mu}_{02}^{C} = (\underline{\mu}_{02}^{C_1} = 0, \ \underline{\mu}_{02}^{C_2} = 0.97, \ \underline{\mu}_{02}^{C_3} = 0.13, \ \underline{\mu}_{02}^{C_4} = 0.89, \ \underline{\mu}_{02}^{C_5} = 0.42),$$

for which the optimization criterion (16) takes the value of $F_2 = 0.1402$.

The obtained null solution allows us to arrange for the genetic search for the solution set $\underline{D}_2(\underline{\mu}_2^{A})$, which is completely determined by the least solution

$$\underline{\mu}^{C} = (\underline{\mu}^{C_1} = 0, \ \underline{\mu}^{C_2} = 0.89, \ \underline{\mu}^{C_3} = 0.13, \ \underline{\mu}^{C_4} = 0.89, \ \underline{\mu}^{C_5} = 0.42)$$

and the two upper solutions $\underline{D}_2^{*} = \{ \overline{\mu}_1^{C}, \overline{\mu}_2^{C} \}$

$$\overline{\mu}_1^{C} = (\overline{\mu}_1^{C_1} = 0, \ \overline{\mu}_1^{C_2} = 0.89, \ \overline{\mu}_1^{C_3} = 0.13, \ \overline{\mu}_1^{C_4} = 1.0, \ \overline{\mu}_1^{C_5} = 0.42);$$

$$\overline{\mu}_2^{C} = (\overline{\mu}_2^{C_1} = 0, \ \overline{\mu}_2^{C_2} = 1.0, \ \overline{\mu}_2^{C_3} = 0.13, \ \overline{\mu}_2^{C_4} = 0.89, \ \overline{\mu}_2^{C_5} = 0.42).$$

Thus, the solution of fuzzy relational equations (28) for the second lower solution $\underline{\mu}_2^{A}$ can be represented in the form of intervals:

$$\underline{D}_2(\underline{\mu}_2^{A}) = \{ \mu^{C_1} = 0, \ \mu^{C_2} = 0.89, \ \mu^{C_3} = 0.13, \ \mu^{C_4} \in [0.89, \ 1.0], \ \mu^{C_5} = 0.42 \}$$

$$\bigcup \{ \mu^{C_1} = 0, \ \mu^{C_2} \in [0.89, \ 1.0], \ \mu^{C_3} = 0.13, \ \mu^{C_4} = 0.89, \ \mu^{C_5} = 0.42 \}. \ (31)$$

The intervals of the values of the input variable for each interval in solutions (29) - (31) can be defined with the help of the membership functions in Figure 6b:

- $x_1^* = 0.75$ or $x_1^* = 1.85$ or $x_1^* = 6.00$ for C_1;

- $x_1^* \in [2.81, 3.25]$ for C_2;

- $x_1^* = 5.27$ or $x_1^* = 4.20$ or $x_1^* = 0$ for C_3;

- $x_2^* \in [0, 0.27]$ for C_4;

- $x_2^* = 2.44$ and $x_2^* = 3.66$ or $x_2^* = 2.35$ and $x_2^* = 3.75$ for C_5.

The restoration of the input set for $y_1^* = 0.20$ and $y_2^* = 0.80$ is shown in Figure 6b, in which the values of the causes and effects significances measures are marked. The comparison of the target and restored level lines for $y_1^* = 0.20$ and $y_2^* = 0.80$ is shown in Figure 7.

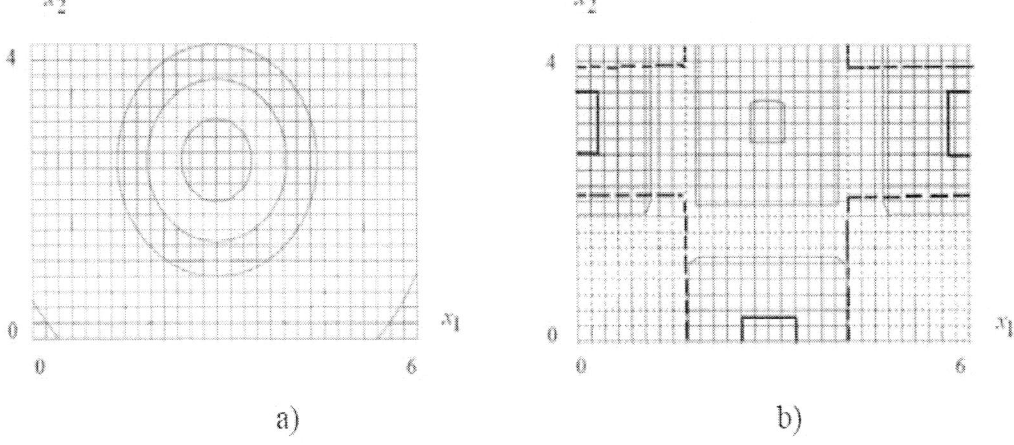

a)   b)

Figure 7. Comparison of the target (a) and restored (b) level lines for $y_1^* = 0.20$ (_____) and $y_2^* = 0.80$ (_ _ _).

7. Example of Fuel Pump Diagnosis

Let us consider the algorithm's performance having the recourse to the example of the fuel pump faults causes diagnosis.

Input parameters are: x_1 – engine speed (30 – 50 r.p.m.); x_2 – inlet pressure (0.02 – 0.15 kg/cm^2); x_3 – drive gear clearance (0.1 – 0.3 mm); x_4 – fuel leakage (0.5 – 2.0 cm^3/h); x_5 – kinematical fuel viscosity ((55–170) $\cdot$ 10^{-6} m^2/c).

Output parameters are: y_1 – productivity (17–22 m^3/h); y_2 – consumed power (2.1–3.0 kw).

Input term-assessments are: c_{11} – engine speed x_1 drop; c_{21} – inlet pressure x_2 drop; c_{31} – increase of clearance x_3, i.e., gear wear-out; c_{41} – increase of leakage x_4, i.e., fuel escape; c_{51} (c_{52}) – low (high) fuel viscosity x_5.

Output term-assessments are: e_{11} (e_{12}) – productivity y_1 fall (rise); e_{21} (e_{22}) – consumed power y_2 drop (rise).

We shall define the set of causes and effects in the following way: $\{C_1, C_2, ..., C_6\} = \{c_{11}, c_{21}, c_{31}, c_{41}, c_{51}, c_{52}\}$; $\{E_1, E_2, ..., E_4\} = \{e_{11}, e_{12}, e_{21}, e_{22}\}$.

«Causes-effects» dependency is described with the help of the following fuzzy relational model

$$(\mu^{A_1}, \mu^{A_2}) * \begin{bmatrix} R_{11} & R_{12} \\ R_{21} & R_{22} \end{bmatrix} = (\mu^{B_1}, \mu^{B_2}), \qquad (32)$$

where $\mu^{A_1} = (\mu^{c_{11}}, \mu^{c_{21}}, \mu^{c_{31}}, \mu^{c_{41}})$, $\mu^{A_2} = (\mu^{c_{51}}, \mu^{c_{52}})$, $\mu^{B_1} = (\mu^{e_{11}}, \mu^{e_{12}})$, $\mu^{B_2} = (\mu^{e_{21}}, \mu^{e_{22}})$.

The resulting expert fuzzy relational matrix takes the following form:

	E_1	E_2	E_3	E_4
C_1	0.44	0.55	0.99	0.12
C_2	0.67	0.11	0.87	0.12
C_3	0.78	0.22	0.55	0.67
C_4	0.67	0.44	0.78	0.12
C_5	0.55	0.67	0.44	0.24
C_6	0.67	0.55	0.87	0.12

$$R = \begin{matrix} c_{11} \\ c_{21} \\ c_{31} \\ c_{41} \\ \\ c_{51} \\ c_{52} \end{matrix} \begin{bmatrix} R_{11} & R_{12} \\ & \\ & \\ & \\ R_{21} & R_{22} \end{bmatrix}$$

For the fuzzy model tuning we used the results of diagnosis for 200 pumps. The results of the fuzzy model tuning are given in Table 8 and Figure 8.

**Table 8. Parameters of the membership functions for the causes
and effects fuzzy terms after tuning**

Parameter	Fuzzy terms					
	c_{11}	c_{21}	c_{31}	c_{41}	c_{51}	c_{52}
β -	32.15	0.031	0.27	1.95	61.56	145.96
σ -	3.75	0.025	0.05	0.48	12.84	28.43

Parameter	Fuzzy terms			
	e_{11}	e_{12}	e_{21}	e_{22}
β -	17.29	21.58	2.11	2.92
σ -	2.72	2.56	0.22	0.25

The system of multidimensional fuzzy relational equations is derived from relation (32). Diagnostic equations after tuning take the following form:

$$\mu^{E1} = [(\mu^{C1} \wedge 0.41) \vee (\mu^{C2} \wedge 0.65) \vee (\mu^{C3} \wedge 0.80) \vee (\mu^{C4} \wedge 0.62)] \wedge [(\mu^{C5} \wedge 0.58) \vee (\mu^{C6} \wedge 0.69)]$$

$$\mu^{E2} = [(\mu^{C1} \wedge 0.59) \vee (\mu^{C2} \wedge 0.09) \vee (\mu^{C3} \wedge 0.23) \vee (\mu^{C4} \wedge 0.45)] \wedge [(\mu^{C5} \wedge 0.69) \vee (\mu^{C6} \wedge 0.50)]$$

$$\mu^{E3} = [(\mu^{C1} \wedge 0.92) \vee (\mu^{C2} \wedge 0.88) \vee (\mu^{C3} \wedge 0.50) \vee (\mu^{C4} \wedge 0.79)] \wedge [(\mu^{C5} \wedge 0.46) \vee (\mu^{C6} \wedge 0.87)]$$

$$\mu^{E4} = [(\mu^{C1} \wedge 0.09) \vee (\mu^{C2} \wedge 0.09) \vee (\mu^{C3} \wedge 0.68) \vee (\mu^{C4} \wedge 0.11)] \wedge [(\mu^{C5} \wedge 0.23) \vee (\mu^{C6} \wedge 0.07)]$$

$$(33)$$

Let us represent the vector of the observed parameters for a specific pump:

$$Y^* = (y_1^* = 19.3 \text{ m}^3/\text{h}; \ y_2^* = 2.23 \text{ kw}).$$

The measures of the effects significances for these values can be defined with the help of the membership functions in Figure 8b:

$$\mu^E = (\mu^{E1}(y_1^*) = 0.65; \ \mu^{E2}(y_1^*) = 0.56; \ \mu^{E3}(y_2^*) = 0.77; \ \mu^{E4}(y_2^*) = 0.12).$$

The genetic algorithm yields a null solution of the system (33)

$$\mu_0^C = (\mu_0^{C1} = 0.62, \ \mu_0^{C2} = 0.77, \ \mu_0^{C3} = 0.12, \ \mu_0^{C4} = 0.39, \ \mu_0^{C5} = 0.56, \ \mu_0^{C6} = 0.88),$$

for which the optimization criterion (15) takes the value of $F_1 = 0.0000$.

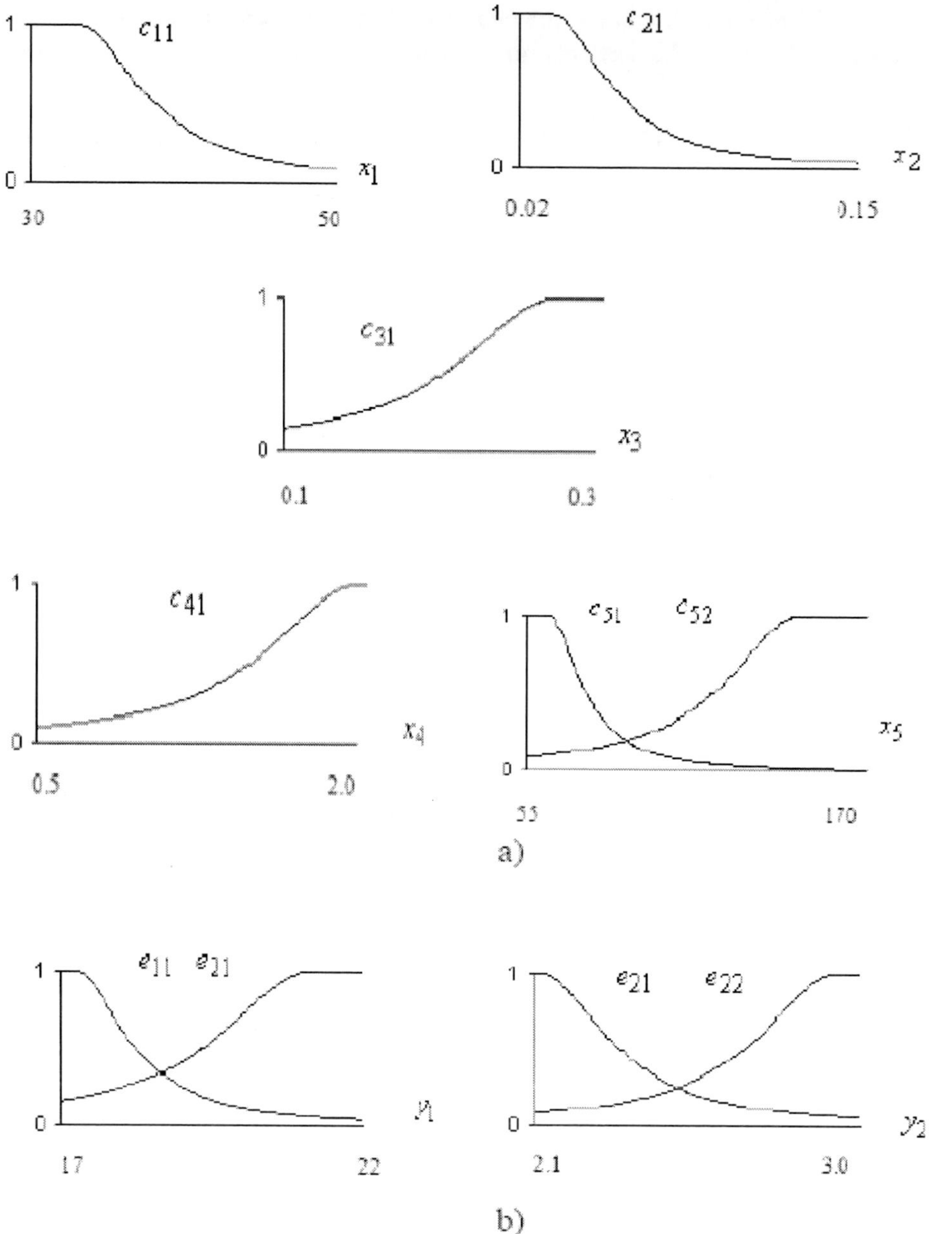

Figure 8. Membership functions of the causes (a) and effects (b) fuzzy terms after tuning.

The obtained null solution allows us to arrange for the genetic search for the solution set $S(R, \mu^E)$, which is completely determined by the four upper solutions

$$\overline{S}^* = \{ \overline{\mu}_1^C, \overline{\mu}_2^C, ..., \overline{\mu}_4^C \}$$

$$\overline{\mu}_1^C = (\overline{\mu}_1^{C_1} = 0.77, \ \overline{\mu}_1^{C_2} = 0.77, \ \overline{\mu}_1^{C_3} = 0.12, \ \overline{\mu}_1^{C_4} = 0.77, \ \overline{\mu}_1^{C_5} = 0.56, \ \overline{\mu}_1^{C_6} = 1.0)$$

$$\overline{\mu}_2^C =(\ \overline{\mu}_2^{C_1}=0.56,\ \overline{\mu}_2^{C_2}=0.77,\ \overline{\mu}_2^{C_3}=0.12,\ \overline{\mu}_2^{C_4}=0.77,\ \overline{\mu}_2^{C_5}=1.0,\ \overline{\mu}_2^{C_6}=1.0)$$

$$\overline{\mu}_3^C =(\ \overline{\mu}_3^{C_1}=1.00,\ \overline{\mu}_3^{C_2}=1.0,\ \overline{\mu}_3^{C_3}=0.12,\ \overline{\mu}_3^{C_4}=1.0,\ \overline{\mu}_3^{C_5}=0.56,\ \overline{\mu}_3^{C_6}=0.77)$$

$$\overline{\mu}_4^C =(\ \overline{\mu}_4^{C_1}=0.56,\ \overline{\mu}_4^{C_2}=1.0,\ \overline{\mu}_4^{C_3}=0.12,\ \overline{\mu}_4^{C_4}=1.0,\ \overline{\mu}_4^{C_5}=1.0,\ \overline{\mu}_4^{C_6}=0.77)$$

and the two lower solutions $\overline{S}^* = \{\underline{\mu}_1^C, \underline{\mu}_2^C\}$

$$\underline{\mu}_1^C =(\ \underline{\mu}_1^{C_1}=0.56,\ \underline{\mu}_1^{C_2}=0.77,\ \underline{\mu}_1^{C_3}=0.12,\ \underline{\mu}_1^{C_4}=0,\ \underline{\mu}_1^{C_5}=0.56,\ \underline{\mu}_1^{C_6}=0.77);$$

$$\underline{\mu}_2^C =(\ \underline{\mu}_2^{C_1}=0.56,\ \underline{\mu}_2^{C_2}=0.65,\ \underline{\mu}_2^{C_3}=0.12,\ \underline{\mu}_2^{C_4}=0.77,\ \underline{\mu}_2^{C_5}=0.56,\ \underline{\mu}_2^{C_6}=0.77).$$

Thus, the solution set of fuzzy relational equations (33) contains the four subsets $D_1(R,\mu^E)$, $D_2(R,\mu^E)$,..., $D_4(R,\mu^E)$, each of which can be represented in the form of intervals:

$$D_1=\{\mu^{C_1}\in[0.56,\ 0.77];\ \mu^{C_2}=0.77;\ \mu^{C_3}=0.12;\ \mu^{C_4}\in[0,\ 0.77];\ \mu^{C_5}=0.56;\ \mu^{C_6}\in[0.77,\ 1.0]\}$$

$$\cup\{\mu^{C_1}\in[0.56,\ 0.77];\ \mu^{C_2}\in[0.65,\ 0.77];\ \mu^{C_3}=0.12;\ \mu^{C_4}=0.77;\ \mu^{C_5}=0.56;\ \mu^{C_6}\in[0.77,\ 1.0]\},\tag{34}$$

$$D_2=\{\mu^{C_1}=0.56;\ \mu^{C_2}=0.77;\ \mu^{C_3}=0.12;\ \mu^{C_4}\in[0,\ 0.77];\ \mu^{C_5}\in[0.56,\ 1.0];\ \mu^{C_6}\in[0.77,\ 1.0]\}$$

$$\cup\{\mu^{C_1}=0.56;\ \mu^{C_2}\in[0.65,\ 0.77];\ \mu^{C_3}=0.12;\ \mu^{C_4}=0.77;\ \mu^{C_5}\in[0.56,\ 1.0];\ \mu^{C_6}\in[0.77,\ 1.0]\},\tag{35}$$

$$D_3=\{\mu^{C_1}\in[0.56,\ 1.0];\ \mu^{C_2}\in[0.77,\ 1.0];\ \mu^{C_3}=0.12;\ \mu^{C_4}\in[0,\ 1.0];\ \mu^{C_5}=0.56;\ \mu^{C_6}=0.77\}$$

$$\cup \{ \mu^{C_1} \in [0.56,\ 1.0];\ \mu^{C_2} \in [0.65,\ 1.0];\ \mu^{C_3} = 0.12;\ \mu^{C_4} \in [0.77,\ 1.0];\ \mu^{C_5} = 0.56;\ \mu^{C_6} = 0.77 \},\tag{36}$$

$$D_4 = \{ \mu^{C_1} = 0.56;\ \mu^{C_2} \in [0.77,\ 1.0];\ \mu^{C_3} = 0.12;\ \mu^{C_4} \in [0,\ 1.0];\ \mu^{C_5} \in [0.56,\ 1.0];\ \mu^{C_6} = 0.77 \}$$

$$\cup \{ \mu^{C_1} = 0.56;\ \mu^{C_2} \in [0.65,\ 1.0];\ \mu^{C_3} = 0.12;\ \mu^{C_4} \in [0.77,\ 1.0];\ \mu^{C_5} \in [0.56,\ 1.0];\ \mu^{C_6} = 0.77 \}.\tag{37}$$

The intervals of the values of the input variables for each interval in solutions (34) - (37) can be defined with the help of the membership functions in Figure 8a:

- $x_1^* \in [34.2, 35.5]$ or $x_1^* \in [30, 35.5]$ r.p.m. for C_1;

- $x_2^* \in [0.045, 0.050]$ or $x_2^* \in [0.02, 0.045]$ or

$x_2^* \in [0.02, 0.050]$ kg/cm^2 for C_2;

- $x_3^* = 0.135$ mm for C_3;

- $x_4^* \in [0.5, 1.69]$ or $x_4^* \in [1.69, 2.0]$ cm^3/h for C_4;

- $x_5^* \in [130, 170] \cdot 10^{-6}$ m^2/c for C_6.

The resulting solution allows the analyst to make the following conclusions. The cause of the observed pump state should be located and identified as the inlet pressure drop to $0.02 - 0.05$ kg/cm^2 with the fuel viscosity being high $((130{-}170) \cdot 10^{-6}$ m^2/c$)$, so that the significance measures of causes C_2 and C_6 are maximal. In addition, the observed state can be the effect of the engine speed drop to $30 - 35$ r.p.m. or the fuel leakage to $1.69 - 2.0$ cm^3/h with the fuel viscosity being high, since the significance measures of causes C_1 and C_4 are sufficiently high. The gear wear-out for the side clearance from 0.1 to 0.135 mm should be excluded, so that the significance measure of cause C_3 is small.

To test the fuzzy model we used the results of diagnosis for 193 pumps with different kinds of faults. The goal was to identify the possible fault causes and evaluate the average percentage of correct diagnosis. The tuning algorithm efficiency characteristics for the testing data are given in Table 9. The fault causes diagnosis started with an average accuracy of 80%

and obtained an accuracy rate of 95% after 10000 iterations of the genetic algorithm (20 min on Intel Core 2 Duo P7350 2.0 GHz).

Table 9. Tuning algorithm efficiency characteristics

Causes (diagnoses)	Number of cases in the data sample	Probability of the correct diagnose	
		before tuning	after tuning
$c_{11}\,c_{51}$	74	60 / 74=0.81	72 / 74=0.97
$c_{21}\,c_{51}$	68	52 / 68=0.76	65 / 68=0.95
$c_{31}\,c_{51}$	92	74 / 92=0.80	88 / 92=0.95
$c_{41}\,c_{51}$	105	91 / 105=0.86	101 / 105=0.96
$c_{11}\,c_{52}$	52	44 / 52=0.84	50 / 52=0.96
$c_{21}\,c_{52}$	70	51 / 70=0.73	68 / 70=0.97
$c_{31}\,c_{52}$	63	50 / 63=0.79	60 / 63=0.95
$c_{41}\,c_{52}$	88	73 / 88=0.83	85 / 88=0.96

8. Example of Hydraulic Elevator Diagnosis

Let us consider the algorithm's performance having the recourse to the example of the hydraulic elevator faults causes diagnosis.

Input parameters of the hydraulic elevator are (variation ranges are indicated in parentheses): x_1 – engine speed (30 – 50 r.p.s); x_2 – inlet pressure (0.02 – 0.15 kg/cm^2); x_3 – feed change gear clearance (0.1 – 0.3 mm).

Output parameters of the hydraulic elevator are: y_1 – productivity (13 – 24 l/min); y_2 – consumed power (2.1 – 3.0 kw); y_3 – suction conduit pressure (0.5 – 1.0 kg/cm^2).

"Causes-effects" interconnection is described with the help of the following system of fuzzy IF-THEN rules:

Rule 1: IF x_1 =I AND x_2 =I AND x_3 =I THEN y_1 =D AND y_2 =I AND y_3 =D;

Rule 2: IF x_1 =D AND x_2 =D AND x_3 =D THEN y_1 =D AND y_2 =D AND y_3 =I;

Rule 3: IF x_1 =I AND x_2 =I AND x_3 =D THEN y_1 =D AND y_2 =D AND y_3 =D;

Rule 4: IF x_1 =I AND x_2 =D AND x_3 =D THEN y_1 =I AND y_2 =D AND y_3 =D;

Rule 5: IF x_1 =D AND x_2 =I AND x_3 =D THEN y_1 =I AND y_2 =D AND y_3 =I, where the total number of the causes and effects consists of: c_{11} *Decrease* (*D*) and c_{12} *Increase* (*I*) for x_1; c_{21} (*D*) and c_{21} (*I*) for x_2; c_{31} (*D*) and c_{32} (*I*) for x_3; e_{11} (*D*) and e_{12} (*I*) for y_1; e_{21} (*D*) and e_{22} (*I*) for y_2; e_{31} (*D*) and e_{32} (*I*) for y_3.

We shall define the set of causes and effects in the following way: $\{C_1, C_2, ..., C_6\} = \{c_{11}, c_{12}, c_{21}, c_{22}, c_{31}, c_{32}\}$; $\{E_1, E_2, ..., E_6\} = \{e_{11}, e_{12}, e_{21}, e_{22}, e_{31}, e_{32}\}$. This fuzzy rule base is modelled by the fuzzy relational matrix presented in Table 10.

For the fuzzy model tuning we used the results of diagnosis for 200 hydraulic elevators. The results of the fuzzy model tuning are given in Table 11 and Figure 9.

Table 10. Fuzzy knowledge matrix

IF inputs				THEN outputs					
	x_1	x_2	x_3	y_1		y_2		y_3	
				D	I	D	I	D	I
A_1	I	I	I	1	0	0	1	1	0
A_2	D	D	D	1	0	1	0	0	1
A_3	I	I	D	1	0	1	0	1	0
A_4	I	D	D	0	1	1	0	1	0
A_5	D	I	D	0	1	1	0	0	1

**Table 11. Parameters of the membership functions for the causes
and effects fuzzy terms after tuning**

Parameter	Fuzzy terms					
	C_1	C_2	C_3	C_4	C_5	C_6
β -	32.15	48.65	0.021	0.144	0.11	0.27
σ -	7.75	6.27	0.054	0.048	0.06	0.08
Parameter	Fuzzy terms					
	E_1	E_2	E_3	E_4	E_5	E_6
β -	13.58	21.43	2.24	2.85	0.53	0.98
σ -	4.76	4.58	0.35	0.17	0.31	0.22

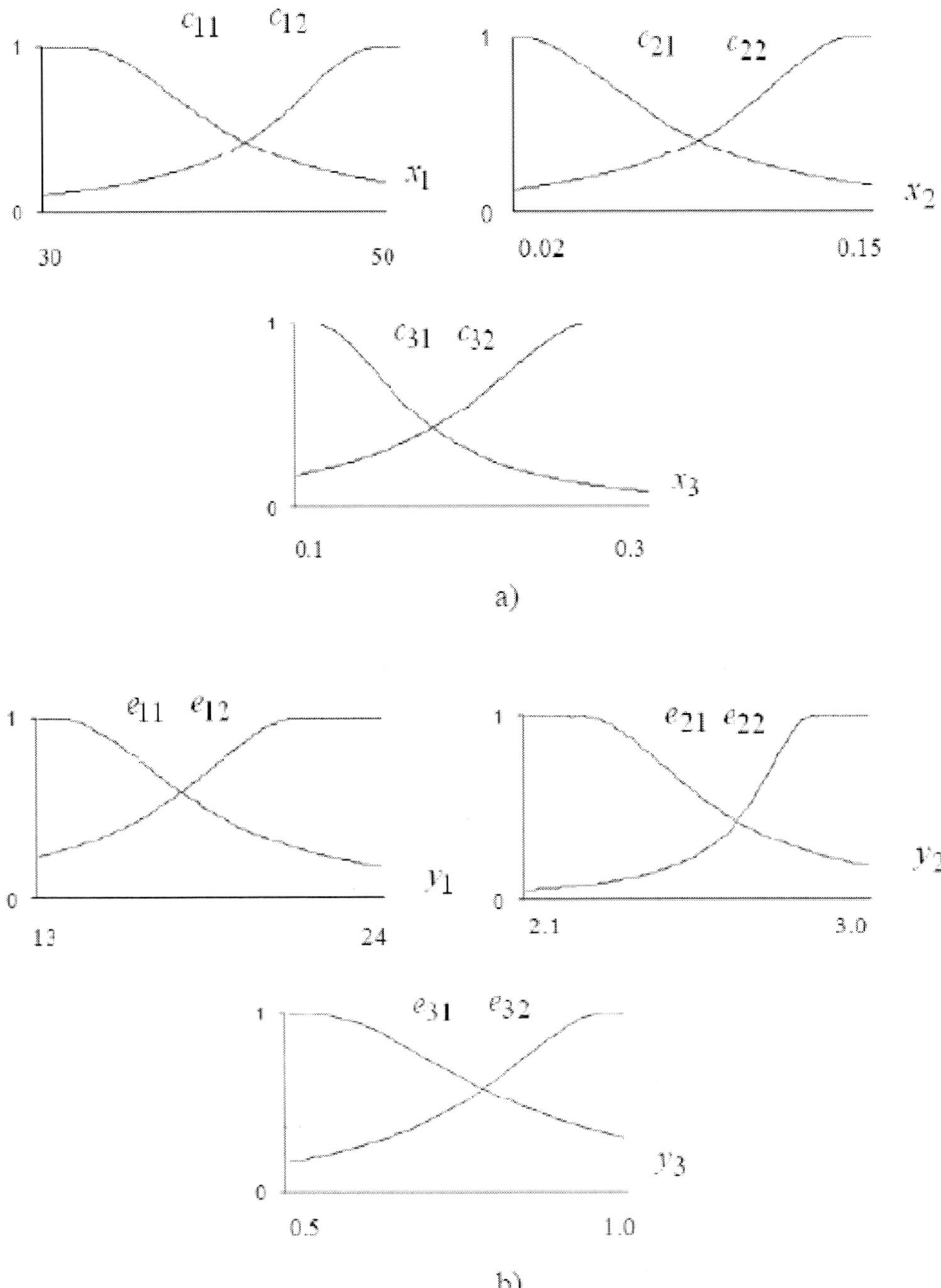

Figure 9. Membership functions of the causes (a) and effects (b) fuzzy terms after tuning.

Diagnostic equations after tuning take the following form:

$$(\mu^{A_1} \wedge 0.97) \vee (\mu^{A_2} \wedge 0.65) \vee (\mu^{A_3} \wedge 0.77) = \mu^{E_1}$$

$$(\mu^{A_4} \wedge 1.00) \vee (\mu^{A_5} \wedge 0.46) = \mu^{E_2}$$

$$(\mu^{A2} \wedge 0.99) \vee (\mu^{A3} \wedge 0.80) \vee (\mu^{A4} \wedge 0.69) \vee (\mu^{A5} \wedge 0.93) = \mu^{E3}$$

$$(\mu^{A1} \wedge 0.96) = \mu^{E4}$$

$$(\mu^{A1} \wedge 0.72) \vee (\mu^{A3} \wedge 0.47) \vee (\mu^{A4} \wedge 0.76) = \mu^{E5}$$

$$(\mu^{A2} \wedge 0.92) \vee (\mu^{A5} \wedge 0.87) = \mu^{E6} \tag{38}$$

where

$$\mu^{C2} \wedge \mu^{C4} \wedge \mu^{C6} = \mu^{A1}$$

$$\mu^{C1} \wedge \mu^{C3} \wedge \mu^{C5} = \mu^{A2}$$

$$\mu^{C2} \wedge \mu^{C4} \wedge \mu^{C5} = \mu^{A3}$$

$$\mu^{C2} \wedge \mu^{C3} \wedge \mu^{C5} = \mu^{A4}$$

$$\mu^{C1} \wedge \mu^{C4} \wedge \mu^{C5} = \mu^{A5}. \tag{39}$$

Let us represent the vector of the observed parameters for a specific elevator:

$$Y^* = (\ y_1^* = 18.80 \text{ l/min}, \ y_2^* = 2.51 \text{ kw}, \ y_3^* = 0.75 \text{ kg/cm}^2).$$

The measures of the effects significances for these values can be defined with the help of the membership functions in Figure 9b:

$$\mu^E = (\ \mu^{E1}(y_1^*) = 0.45; \ \mu^{E2}(y_1^*) = 0.75;$$

$$\mu^{E3}(y_2^*) = 0.63; \ \mu^{E4}(y_2^*) = 0.20;$$

$$\mu^{E5}(y_3^*) = 0.67; \ \mu^{E6}(y_3^*) = 0.48).$$

The genetic algorithm yields a null solution of the system (38)

$$\mu_0^C = (\mu_0^{C1} = 0.46, \ \mu_0^{C2} = 0.69, \ \mu_0^{C3} = 0.75, \ \mu_0^{C4} = 0.25, \ \mu_0^{C5} = 0.92, \ \mu_0^{C6} = 0.20),$$

for which the optimization criterion (16) takes the value of F_2=0.0081.

The null vector of the causes combinations significances measures

$$\mu_0^A = (\mu_0^{A_1} = 0.20, \ \mu_0^{A_2} = 0.46, \ \mu_0^{A_3} = 0.25, \ \mu_0^{A_4} = 0.69, \ \mu_0^{A_5} = 0.25)$$

corresponds to the obtained null solution.

The obtained null solution allows us to arrange for the genetic search for the solution set $S(R, \mu^E)$, which is completely determined by the greatest solution

$$\overline{\mu}^A = (\ \overline{\mu}^{A_1} = 0.20, \ \overline{\mu}^{A_2} = 0.46, \ \overline{\mu}^{A_3} = 0.46, \ \overline{\mu}^{A_4} = 0.69, \ \overline{\mu}^{A_5} = 0.46)$$

and the two lower solutions $S^* = \{\underline{\mu}_1^A, \underline{\mu}_2^A\}$

$$\underline{\mu}_1^A = (\ \underline{\mu}_1^{A_1} = 0.20, \ \underline{\mu}_1^{A_2} = 0.46, \ \underline{\mu}_1^{A_3} = 0, \ \underline{\mu}_1^{A_4} = 0.69, \ \underline{\mu}_1^{A_5} = 0);$$

$$\underline{\mu}_2^A = (\ \underline{\mu}_2^{A_1} = 0.20, \ \underline{\mu}_2^{A_2} = 0, \ \underline{\mu}_2^{A_3} = 0.46, \ \underline{\mu}_2^{A_4} = 0.69, \ \underline{\mu}_2^{A_5} = 0.46).$$

Thus, the solution of fuzzy relational equations (38) can be represented in the form of intervals:

$$S(R, \mu^E) = \{\ \mu^{A_1} = 0.20, \ \mu^{A_2} = 0.46, \ \mu^{A_3} \in [0, \ 0.46], \ \mu^{A_4} = 0.69, \ \mu^{A_5} \in [0,$$

$$0.46]\}$$

$$\cup \{\ \mu^{A_1} = 0.20, \ \mu^{A_2} \in [0, 0.46], \ \mu^{A_3} = 0.46, \ \mu^{A_4} = 0.69, \ \mu^{A_5} = 0.46\}.$$

We next apply the genetic algorithm for solving (39) for the greatest solution $\overline{\mu}^A$ and the two lower solutions $\underline{\mu}_1^A$ and $\underline{\mu}_2^A$.

For the greatest solution $\overline{\mu}^A$, the genetic algorithm yields a null solution of the system (39)

$$\overline{\mu}_0^C = (\overline{\mu}_0^{C_1} = 0.46, \ \overline{\mu}_0^{C_2} = 0.78, \ \overline{\mu}_0^{C_3} = 0.69, \ \overline{\mu}_0^{C_4} = 0.46, \ \overline{\mu}_0^{C_5} = 0.91, \ \overline{\mu}_0^{C_6} = 0.20),$$

for which the optimization criterion (16) takes the value of F_2=0.0081.

The obtained null solution allows us to arrange for the genetic search for the solution set $\overline{D}(\ \overline{\mu}^A)$, which is completely determined by the least solution

$$\underline{\mu}^{C} = (\ \underline{\mu}^{C_1} = 0.46, \ \underline{\mu}^{C_2} = 0.69, \ \underline{\mu}^{C_3} = 0.69, \ \underline{\mu}^{C_4} = 0.46, \ \underline{\mu}^{C_5} = 0.69, \ \underline{\mu}^{C_6} = 0.20)$$

and the three upper solutions $\overline{D}^{*} = \{\overline{\mu}_1^{C}, \overline{\mu}_2^{C}, \overline{\mu}_3^{C}\}$

$$\overline{\mu}_1^{C} = (\ \overline{\mu}_1^{C_1} = 0.46, \ \overline{\mu}_1^{C_2} = 0.69, \ \overline{\mu}_1^{C_3} = 1.0, \ \overline{\mu}_1^{C_4} = 0.46, \ \overline{\mu}_1^{C_5} = 1.0, \ \overline{\mu}_1^{C_6} = 0.20);$$

$$\overline{\mu}_2^{C} = (\ \overline{\mu}_2^{C_1} = 0.46, \ \overline{\mu}_2^{C_2} = 1.0, \ \overline{\mu}_2^{C_3} = 0.69, \ \overline{\mu}_2^{C_4} = 0.46, \ \overline{\mu}_2^{C_5} = 1.0, \ \overline{\mu}_2^{C_6} = 0.20);$$

$$\overline{\mu}_3^{C} = (\ \overline{\mu}_3^{C_1} = 0.46, \ \overline{\mu}_3^{C_2} = 1.0, \ \overline{\mu}_3^{C_3} = 1.0, \ \overline{\mu}_3^{C_4} = 0.46, \ \overline{\mu}_3^{C_5} = 0.69, \ \overline{\mu}_3^{C_6} = 0.20).$$

Thus, the solution of fuzzy relational equations (39) for the greatest solution $\overline{\mu}^{A}$ can be represented in the form of intervals:

$$\overline{D}(\overline{\mu}^{A}) = \{\ \mu^{C_1} = 0.46, \ \mu^{C_2} = 0.69, \ \mu^{C_3} \in [0.69,\ 1.0], \ \mu^{C_4} = 0.46, \ \mu^{C_5} \in [0.69,$$

$1.0], \ \mu^{C_6} = 0.20\}$

$$\bigcup \{\ \mu^{C_1} = 0.46, \ \mu^{C_2} \in [0.69,\ 1.0], \ \mu^{C_3} = 0.69, \ \mu^{C_4} = 0.46, \ \mu^{C_5} \in [0.69,\ 1.0],$$

$\mu^{C_6} = 0.20\}$

$$\bigcup \{\ \mu^{C_1} = 0.46, \ \mu^{C_2} \in [0.69,\ 1.0], \ \mu^{C_3} \in [0.69,\ 1.0], \ \mu^{C_4} = 0.46, \ \mu^{C_5} = 0.69,$$

$$\mu^{C_6} = 0.20\} \tag{40}$$

For the first lower solution $\underline{\mu}_1^{A}$, the genetic algorithm yields a null solution of the system (39)

$$\underline{\mu}_{01}^{C} = (\underline{\mu}_{01}^{C_1} = 0.46, \underline{\mu}_{01}^{C_2} = 0.92, \underline{\mu}_{01}^{C_3} = 0.86, \underline{\mu}_{01}^{C_4} = 0.10, \underline{\mu}_{01}^{C_5} = 0.69, \underline{\mu}_{01}^{C_6} = 0.10),$$

for which the optimization criterion (16) takes the value of $F_2 = 0.0181$.

The obtained null solution allows us to arrange for the genetic search for the solution set $\underline{D}_1(\underline{\mu}_1^{A})$, which is completely determined by the least solution

$$\underline{\mu}^{C} = (\ \underline{\mu}^{C_1} = 0.46, \ \underline{\mu}^{C_2} = 0.69, \ \underline{\mu}^{C_3} = 0.69, \ \underline{\mu}^{C_4} = 0.10, \ \underline{\mu}^{C_5} = 0.69, \ \underline{\mu}^{C_6} = 0.10)$$

and the three upper solutions $\overline{D}_1^* = \{\overline{\mu}_1^C, \overline{\mu}_2^C, \overline{\mu}_3^C\}$

$$\overline{\mu}_1^C = (\overline{\mu}_1^{C_1} = 0.46, \ \overline{\mu}_1^{C_2} = 0.69, \ \overline{\mu}_1^{C_3} = 1.0, \ \overline{\mu}_1^{C_4} = 0.10, \ \overline{\mu}_1^{C_5} = 1.0, \ \overline{\mu}_1^{C_6} = 0.10);$$

$$\overline{\mu}_2^C = (\overline{\mu}_2^{C_1} = 0.46, \ \overline{\mu}_2^{C_2} = 1.0, \ \overline{\mu}_2^{C_3} = 0.69, \ \overline{\mu}_2^{C_4} = 0.10, \ \overline{\mu}_2^{C_5} = 1.0, \ \overline{\mu}_2^{C_6} = 0.10);$$

$$\overline{\mu}_3^C = (\overline{\mu}_3^{C_1} = 0.46, \ \overline{\mu}_3^{C_2} = 1.0, \ \overline{\mu}_3^{C_3} = 1.0, \ \overline{\mu}_3^{C_4} = 0.10, \ \overline{\mu}_3^{C_5} = 0.69, \ \overline{\mu}_3^{C_6} = 0.10).$$

Thus, the solution of fuzzy relational equations (39) for the first lower solution $\underline{\mu}_1^A$ can be represented in the form of intervals:

$$\underline{D}_1(\underline{\mu}_1^A) = \{\ \mu^{C_1} = 0.46, \ \mu^{C_2} = 0.69, \ \mu^{C_3} \in [0.69, \ 1.0], \ \mu^{C_4} = 0.10, \ \mu^{C_5} \in [0.69,$$
$$1.0], \ \mu^{C_6} = 0.10\}$$

$$\cup \{\ \mu^{C_1} = 0.46, \ \mu^{C_2} \in [0.69, \ 1.0], \ \mu^{C_3} = 0.69, \ \mu^{C_4} = 0.10, \ \mu^{C_5} \in [0.69, \ 1.0],$$
$$\mu^{C_6} = 0.10\}$$

$$\cup \{\ \mu^{C_1} = 0.46, \ \mu^{C_2} \in [0.69, \ 1.0], \ \mu^{C_3} \in [0.69, \ 1.0], \ \mu^{C_4} = 0.10, \ \mu^{C_5} = 0.69,$$
$$\mu^{C_6} = 0.10\}. \tag{41}$$

For the second lower solution $\underline{\mu}_2^A$, the genetic algorithm yields a null solution of the system (39)

$$\underline{\mu}_{02}^C = (\underline{\mu}_{02}^{C_1} = 0.23, \underline{\mu}_{02}^{C_2} = 0.69, \underline{\mu}_{02}^{C_3} = 0.83, \underline{\mu}_{02}^{C_4} = 0.46, \underline{\mu}_{02}^{C_5} = 0.97, \underline{\mu}_{02}^{C_6} = 0.20),$$

for which the optimization criterion (16) takes the value of $F_2 = 0.0702$.

The obtained null solution allows us to arrange for the genetic search for the solution set $\underline{D}_2(\underline{\mu}_2^A)$, which is completely determined by the least solution

$$\underline{\mu}^C = (\ \underline{\mu}^{C_1} = 0.23, \ \underline{\mu}^{C_2} = 0.69, \ \underline{\mu}^{C_3} = 0.69, \ \underline{\mu}^{C_4} = 0.46, \ \underline{\mu}^{C_5} = 0.69, \ \underline{\mu}^{C_6} = 0.20)$$

and the three upper solutions $\underline{D}_2^* = \{\overline{\mu}_1^C, \overline{\mu}_2^C, \overline{\mu}_3^C\}$

$$\overline{\mu}_1^C = (\overline{\mu}_1^{C_1} = 0.23, \ \overline{\mu}_1^{C_2} = 0.69, \ \overline{\mu}_1^{C_3} = 1.0, \ \overline{\mu}_1^{C_4} = 0.46, \ \overline{\mu}_1^{C_5} = 1.0, \ \overline{\mu}_1^{C_6} = 0.20);$$

$$\overline{\mu}_2^C = (\overline{\mu}_2^{C_1} = 0.23, \ \overline{\mu}_2^{C_2} = 1.0, \ \overline{\mu}_2^{C_3} = 0.69, \ \overline{\mu}_2^{C_4} = 0.46, \ \overline{\mu}_2^{C_5} = 1.0, \ \overline{\mu}_2^{C_6} = 0.20);$$

$$\overline{\mu}_3^C = (\overline{\mu}_3^{C_1} = 0.23, \ \overline{\mu}_3^{C_2} = 1.0, \ \overline{\mu}_3^{C_3} = 1.0, \ \overline{\mu}_3^{C_4} = 0.46, \ \overline{\mu}_3^{C_5} = 0.69, \ \overline{\mu}_3^{C_6} = 0.20).$$

Thus, the solution of fuzzy relational equations (39) for the second lower solution $\underline{\mu}_2^A$ can be represented in the form of intervals:

$$\underline{D}_2(\underline{\mu}_2^A) = \{ \mu^{C_1} = 0.23, \ \mu^{C_2} = 0.69, \ \mu^{C_3} \in [0.69, \ 1.0], \ \mu^{C_4} = 0.46, \ \mu^{C_5} \in [0.69,$$

$$1.0], \ \mu^{C_6} = 0.20\}$$

$$\bigcup \{ \mu^{C_1} = 0.23, \ \mu^{C_2} \in [0.69, \ 1.0], \ \mu^{C_3} = 0.69, \ \mu^{C_4} = 0.46, \ \mu^{C_5} \in [0.69, \ 1.0],$$

$$\mu^{C_6} = 0.20\}$$

$$\bigcup \{ \mu^{C_1} = 0.23, \ \mu^{C_2} \in [0.69, \ 1.0], \ \mu^{C_3} \in [0.69, \ 1.0], \ \mu^{C_4} = 0.46, \ \mu^{C_5} = 0.69,$$

$$\mu^{C_6} = 0.20\}. \tag{42}$$

Following the resulting solutions (40) - (42), the causes C_2, C_3 and C_5 are the causes of the observed elevator state, so that $\mu^{C_2} > \mu^{C_1}$, $\mu^{C_3} > \mu^{C_4}$, $\mu^{C_5} > \mu^{C_6}$. The intervals of the values of the input variables for these causes can be defined with the help of the membership functions in Figure 9a:

- $x_1^* \in [44.4, 50.0]$ r.p.s for C_2;

- $x_2^* \in [0.020, 0.057]$ kg/cm^2 for C_3;

- $x_3^* \in [0.100, 0.150]$ mm for C_5.

Thus, the causes of the observed elevator state should be located and identified as the increase of the engine speed to 45-50 r.p.s, the decrease of the inlet pressure to $0.020 - 0.057$ kg/cm^2, and the decrease of the feed change gear clearance to 100-150 mk.

Table 12. Tuning algorithm efficiency characteristics

Causes (diagnoses)	Number of cases in the data sample	Probability of the correct diagnose	
		before tuning	after tuning
c_{11}	104	86 / 104=0.82	101 / 104=0.97
c_{12}	88	67 / 88=0.76	84 / 88=0.95
c_{21}	92	74 / 92=0.80	88 / 92=0.95
c_{22}	100	70 / 100=0.70	97 / 100=0.97
c_{31}	122	103 / 122=0.84	117 / 122=0.96
c_{32}	70	51 / 70=0.73	68 / 70=0.97

To test the fuzzy model we used the results of diagnosis for 192 elevators with different kinds of faults. The tuning algorithm efficiency characteristics for the testing data are given in Table 12. The fault causes diagnosis started with an average accuracy of 77.5% and obtained an accuracy rate of 96% after 10000 iterations of the genetic algorithm (20 min on Intel Core 2 Duo P7350 2.0 GHz).

Conclusion

In the cases, when domain experts are involved in developing cause-effect connections, the dependency between unobserved and observed parameters can be modelled using the means of fuzzy sets theory: fuzzy relations and fuzzy IF-THEN rules. Application of expert fuzzy relational rules to restore and identify the causes by observed effects requires solving fuzzy relational equations. We suggest some procedures of numerical solution of the multidimensional fuzzy relational equations using genetic algorithms. Although the work presented here shows good practical results, some future investigations are still needed. While the theoretical foundations of the fuzzy relational equations are well developed, they still call for more efficient and diversified schemes of solution finding. The issue of adaptation of the resulting solution while the observed effects are changing remains unresolved. The genetically guided global optimization should be augmented by more refined gradient-based adaptation mechanisms [5]. Such an adaptive approach envisages the development of a hybrid genetic and neuro algorithm for solving multidimensional fuzzy relational equations. The approach proposed can find application not only in engineering but also in medicine, economics, military affairs and other domains, from which the necessity for interpreting the experimental observations arises.

References

[1] Zadeh, L. (1976). *The Concept of Linguistic Variable and It's Application to Approximate Decision Making.* Moscow: Mir.

[2] Yager, R. R., and Filev, D. P. (1994). *Essentials of Fuzzy Modelling and Control.* John Willey and Sons, New York.

[3] Peeva, K., and Kyosev, Y. (2004). *Fuzzy Relational Calculus Theory, Applications and Software.* World Scientific Publishing Company. 292 p. CD-ROM http://math works.net.

[4] Rotshtein, A., and Rakytyanska, H. *Fuzzy genetic object identification: multiple-inputs multiple-outputs case.* In: Hippe, Z. S., Kulikowski, J. L., Mroczek, T. (Eds.) *Human-Computer Systems Interaction. Backgrounds and Applications. Part II.* Advances in Soft Computing, Springer, 2011 (to be appeared).

[5] Rotshtein, A., and Rakytyanska, H. (2009). Diagnosis based on fuzzy IF-THEN rules and genetic algorithms. In: Hippe, Z.S., and Kulikowski, J. L. (Eds.) *Human-Computer Systems Interaction. Backgrounds and Applications. Part I.* Advances in Soft Computing, Springer, Vol. 60, pp. 541-556.

[6] Sanchez, E. (1976). Resolution of composite fuzzy relation equations. *Information and Control,* vol. 30 (1), pp. 38-48.

[7] Miyakoshi, M., and Shimbo, M. (1986). Lower solutions of systems of fuzzy equations. *Fuzzy Sets and Systems,* vol. 19, pp. 37-46.

[8] Di Nola, A., Sessa, S., Pedrycz, W., and Sanchez, E. (1989). *Fuzzy Relation Equations and Their Applications to Knowledge Engineering,* Dordrecht: Kluwer Academic Press.

[9] De Baets, B. (2000). Analytical solution methods for fuzzy relational equations. In: Dubois, D., and Prade, H. (Eds.) *Fundamentals of Fuzzy Sets, The Handbooks of Fuzzy Sets Series,* vol. 1, Kluwer Academic Publishers, pp. 291-340.

[10] Gottwald, S., and Pedrycz, W. (1986) Solvability of fuzzy relational equations and manipulation of fuzzy data. *Fuzzy Sets and Systems,* vol. 18 (1), pp. 45-65.

[11] Neundorf, D., and Bohm, R. (1996). Solvability criteria for systems of fuzzy relation equations. *Fuzzy Sets and Systems,* vol. 80 (3), pp. 345-352.

[12] Gottwald, S., and Perfilieva, I. (2003). Solvability and approximate solvability of fuzzy relation equations. *Intern. J. General Systems,* vol. 32 (4), pp. 361-372.

[13] Higashi, M., and Klir, G. J. (1984). Identification of fuzzy relation systems. *IEEE Transactions Systems, Man, and Cybernetics,* vol. 14, pp. 349-355.

[14] Pedrycz, W. (1984). An identification algorithm in fuzzy relational systems, *Fuzzy Sets and Systems,* vol. 13, pp. 153-167.

[15] Rotshtein, A. (1998). Design and tuning of fuzzy rule-based systems for medical diagnosis. In: Teodorescu, N.-H., Kandel, A., and Gain, L. (Eds.) *Fuzzy and Neuro-Fuzzy Systems in Medicine,* Boca-Raton: CRC Press, pp. 243-289.

[16] Rotshtein, A., and Rakytyanska, H. (2007). Fuzzy relation-based diagnosis. *Automation and Remote Control,* 68 (12), pp. 2198-2213.

[17] Rotshtein, A., and Rakytyanska, H. (2008). Diagnosis problem solving using fuzzy relations. *IEEE Transactions on Fuzzy Systems.* Vol. 16 (3), pp. 664 – 675.

[18] Rotshtein, A., and Rakytyanska, H. (2009) Adaptive diagnostic system based on fuzzy relations. *Cybernetics and Systems Analysis*, 45 (4), pp. 623-637.

[19] Rotshtein, A., and Rakytyanska, H. (2001). Genetic algorithm of fuzzy logic equations solving in expert systems of diagnosis. In: Monostori, L., Vancza, J., and Ali, M. (Eds.) *Engineering of Intelligent Systems, Lecture Notes in Computer Science,* Springer Verlag, pp. 139-148.

[20] Rotshtein, A., Posner, M., and Rakytyanska, H. (2006). Cause and effect analysis by fuzzy relational equations and a genetic algorithm. *Reliability Engineering and System Safety*, vol. 91(9), pp. 1095-1101.

[21] Rotshtein, A. (1997). Modification of Saaty method for the construction of fuzzy set membership functions, In Proc. FUZZY'97 Intern. Conf. on Fuzzy Logic and Its Applications, Zichron Yaakov, Israel, pp. 125-130.

[22] Gen, M., and Cheng, R. (1997). *Genetic Algorithms and Engineering Design.* John Wiley and Sons.

[23] Cordon, O., Herrera, F., Hoffmann F., and Magdalena, L. (2001). *Genetic Fuzzy Systems. Evolutionary Tuning and Learning of Fuzzy Knowledge Bases.* World Scientific. 492 p.

In: Genetic Diagnoses
Editor: R. J. Sarma, pp. 99-114

ISBN: 978-1-61324-866-9
© 2012 Nova Science Publishers, Inc.

Chapter IV

Molecular Diagnosis
of Spinocerbellar Ataxias

*Laia Rodriguez-Revenga, Irene Madrigal, Maria Isabel Alvarez
and Montserrat Milà** *
Biochemistry and Molecular Genetics Department.
Hospital Clínic. Barcelona, Spain
CIBER de Enfermedades Raras (CIBERER), Barcelona, Spain

Abstract

Spinocerebelar ataxias (SCA) are a clinical and genetic heterogeneous group of
neurodegenerative disease, affecting 1-4 cases per 100,000 individuals and presenting a
wide geographic variation. SCAs are characterized by progressive degeneration of
cerebellum and its afferent and efferent connections. The main clinical symptom is the
presence of gait and limb ataxia, and most of them are inherited as a dominant trait.
Based in clinical manifestations SCAs can be classified in four groups, but based on the
genetic molecular defects, there are around 30 different SCAs, nevertheless some patients
remain without a molecular diagnosis. Although the first approach is clinic, molecular
diagnosis becomes essential, providing an accurate diagnostic and making possible
genetic counselling, presymptomatic testing and prenatal diagnosis. It is important to
highlight the fact that while a positive test establish a diagnostic, a negative one does not
discard it. The main molecular defect found in SCAs is a dynamic mutation. In most of
the cases, it corresponds to a CAG expansion in the coding region of the gene, giving a
polyQ protein and sharing the particularities of this kind of mutations: late onset disease
with variable expression, complete penetrance, anticipation and being the paternal
transmission more severe than the maternal one. SCA alleles are classified in four
categories normal, intermediate, reduced penetrance and mutated. Individuals carrying
the mutation synthesize an aberrant protein, with an acquired gain of function that
damages particular group of cells and that leads to cell death.

---

* Tel. +34 93 227 54 00 Ext.2784, Fax. +34 93 451 52 72, e-mail: mmila@clinic.ub.es.

There are other cases of ataxia, such as FXTAS, in which the molecular defect is also a dynamic mutation but localized in the non-coding region of the gene. Finally there are also ataxias caused by rare conventional mutations (missense, deletion, insertion or duplication). An example is SCA11, which is caused by mutations in *TTBK2* gene.

In this chapter we present our experience in molecular diagnosis of spinocerebellar ataxias. We have performed molecular diagnoses of SCA in more than 500 unrelated patients, including confirmation of clinical suspicion, prenatal and presymptomatic diagnosis for SCA1, SCA2, SCA3, SCA6, SCA7, SCA8, DRPLA and FXTAS. On the basis of our results, we can conclude that the most common ataxia in the Spanish population is SCA3 (Machado Joseph) followed by SCA2. Finally and due to the fact that we are a reference centre for fragile X syndrome, we have a wide experience on FXTAS.

Introduction

Spinocerebellar ataxias (SCAs) are a clinically and genetically heterogeneous complex group of neurodegenerative diseases, affecting cerebellum and spinal cord. SCAs are clinically characterized by a progressive ataxia, alterations of the balance and motor coordination due to cerebellum pathology and/or its connections. Additional clinical sings can also be present, including ophthalmoplegia, pyramidal and extrapyramidal signs, dementia, pigmentary retinopathy and peripheral neuropathy. Neuropathologically it is observed atrophy and neuronal loss in areas and grey centers of the cerebral cortex (Purkinje cells), brainstem, basal nodes and at the spinal olivopontocerebellar atrophy (OPCA,) [1]. SCAs are late onset diseases, starting between 30 and 40 years, although early onset in childhood and onset in later decades after 60 years have also been reported. The prognosis is variable depending on the ataxia subtype but it usually causes death 20 years after the disease onset.

In 1983 Harding reported the first classification of the Autosomal Dominant Cerebellar Ataxia (ADCA) based on clinical aspects, differentiating four groups (I-IV) (Table1) [2][3] (Harding, 1983, revised in Harding1993). Although this classification does not have into consideration genetic aspects, it is still useful as a guideline in clinical practice and to address genetic tests for diagnosis. Based on ADCA classification we differentiate: ADCA I, which encompasses the most, prevalent and heterogeneous group. Besides the ataxia, other additional features such as, ophthalmoplegia, pyramidal and extrapyramidal signs, cognitive impairment, tremor, rigidity, spasticity, and neuronopathy can be present. ADCAII that collects cases presenting with retinal (Retinitis Pigmentosa) and macular degeneration, with optical atrophy and loss of visual acuity or blindness. ADCAIII, which concerns the group where the degenerative process is limited to the cerebellum ("pure ataxia"), and ADCAIV, that refers to those ataxias where, and in addition to the cerebellar ataxia, deafness and myoclonia may be also present.

Regarding genetic aspects, SCAs are very heterogeneous diseases. While the clinical classification only differentiates four groups, genetic classification identifies more than 30 different subtypes with little clinical variations between them. Nowadays more than 30 different loci have been described, corresponding to different dominant forms of SCA. It is possible to classify one of the genetic subtypes into different ADCA groups. For example SCA8 can be clinically included in ADCA I or III, and Dentarubropallidoluisina (DRPLA) in ADCA I and IV.

Table 1. Spinocerebellar ataxias: Harding's classification

Group	Clinical Manifestation	Genotype
ADCA I	- Ataxia - Degeneration of other neuronal systems and/or ophthalmoplegia	SCA1,2,3,4,8,12,13,17,18,19,20,21,23,24,2,5,27,28 and DRPLA
ADCA II	- Ataxia - Degeneration of retina	SCA7
ADCA III	- Pure cerebellar atàxia	SCA 5,6,8,10,11,12,14,15,16,26,29 and 30
ADCA IV	- Ataxia - Myoclonus - Deafness	SCA 14,19,24 and DRPLA

ADCA: Autosomal Dominant Cerebellar Ataxia.

As far as epidemiologic data, SCAs are rare diseases with a prevalence of 1-4/100,000, and showing large geographic variation, probably due to a founder effect in the mutation of the different genes (reviewed in [4]). The most common form in Europe is SCA 3 with a general prevalence of 1-3/100,000 individuals. However, this value increases to 17.8 for each 100,000 individuals in Portugal, and in Azores islands it reaches 1 each 4,000 inhabitant. SCA2 is the most prevalent in Cuba with an estimated prevalence of 40/100,000 individuals. For this reason and in order to provide an accurate genetic counseling it is always important to know the ethnic background of the affected individual.

According to the molecular defect that causes ataxia we can establish three groups. In the first one, the mutation responsible of the defect is an expansion of a CGA triplet in the coding region of the gene. As the expansion is transcribed and translated, it leads to a protein which contains an elevate number of glutamines and for this reason this group of diseases are also called polyglutamine (polyQ) expansion diseases. This group includes SCA1 SCA2, SCA3, SCA6, SCA7, SCA17 and DRPLA. Conventional mutations in these genes do not cause ataxia. In the second group, SCAs are also caused by expansions of three or more nucleotides but the repeats are localized in a non-coding region of the gene. Therefore the expansion is neither transcribed nor translated. This group includes FXTAS, SCA8, SCA10, SCA12 and SCA31. As in the previous group, conventional mutations in these genes do not cause ataxia. Finally, the last group is caused by conventional mutations, such as point mutations, deletions etc…in different genes like SCA5, SCA11, SCA13 etc…

The aim of this chapter is to review the genetic aspects of the different subtypes of SCAs. Moreover and due to the fact that FXTAS has been recently included in the SCAs study protocols, a specific review of this syndrome is included.

Polyglutamine (polyQ) Expansions (Coding Expansion SCAs)

This subtype of SCAs is inherited as a dominant trait, and therefore they are called Autosomal Dominat Cerebelar Ataxias (ADCAs). The molecular defect is a dynamic mutation, an expansion of a CAG repeat in the coding region of the responsible gene. The number of CAG repeats in general population is polymorphic until a threshold above which, the protein gains a new function and becomes cytotoxic. This mutation induces the synthesis of a protein called ataxin which presents a glutamine excess that confers a gain of function and a toxic effect causing neurodegeneration. This protein leads to the formation of intranuclear inclusions in different types of cerebellar cells disturbing the ubiquitin proteasome pathway as well as the chaperones pathway [5]. In all these cases, disease is caused by a mutated protein that leads to the failure of the normal neuronal function and also to the neurodegeneration of a specific group of neurons which are different depending on the subtype of SCAs.

Dynamic mutations exhibit a set of characteristic hallmarks: complete penetrance, variable expressivity and anticipation. Dominat cerebelar ataxias exhibit complete penetrance, and therefore all individuals carrying a mutated allele will develop the disease. In the general population, three different categories of SCA alleles can be found (Table 2). While normal alleles are never associated with the development of the disease, mutated alleles will always develop the disease and have a 50% risk of transmission. Finally, individuals who carry permutated alleles or alleles with reduced penetrance will probably not develop the disease but are at risk of transmitting a mutated allele to the next generation. Therefore, genetic counseling is necessary for individuals with the CAG repeat in the premutation range.

Variable expressivity is also common in SCAs, which allows establishing a positive correlation between the CAG number and clinical manifestations. For example, in SCA7 pathological alleles are over 40 repeats but only individuals with alleles over 60 CAGs show vision loss. Moreover, the CAG repeat number might determine the age of onset. It has been observed that early onset forms correspond to individuals carrying large expansions, while late onset have been reported in individuals with small expansions that are close to normal alleles. Nevertheless, it is known that variability in age of onset and clinical severity is not completely attributable to the CAG repeat expansion and others factors might have a potential role.

Another characteristic phenomenon observed in SCAs is the anticipation. It is possible to track how the CAG repeat number increases thought the different generations of a family and to correlate it with a decrease in the age of onset.

Finally, in some subtypes of SCAs an imprinting effect is also observed. In general when the mutation is transmitted by the father the expansions tend to suffer a larger expansion increase than when is transmitted by the mother.

Table 2 summarizes the different subtypes of SCAs that belong to this polyglutamine expansions group.

Non-Coding Expansion SCAs

Some SCAs are also caused by triplet expansions but the repeat tract is not a GAC and it is not placed in the coding region of the genes. These cases are less common and include SCA8, SCA10, SCA 17 and FXTAS (Table 3). They do not share the same characteristics described for those caused by an expansion in the coding region. Nevertheless, in some cases it can also be established a correlation between age of onset and the size of the repeat.

However, and contrary to what happen with polyQ expansion diseases, early onset forms are associated with low repeat range expansions, as described for example in SCA10 [6].

The molecular mechanism by which the disease is caused is different for each subtype of SCA. In general when expansions occur outside the coding region, they tend to be larger than when they occur in a coding region.

In such cases the toxic effect is produced by other causes rather than by the abnormal protein. For example in SCA10, chromatin abnormalities are responsible of the disease [7], and in FXTAS the abnormal RNA with an abnormal CGG repeat tract is causative of the toxic effect (reviewed in Hagerman and Hagerman, 2004)[8].

As for SCA8, there is controversy regarding if it should still be included in the routine SCA study protocol. There are few cases of individuals carrying expansions in the pathological range (>100 CTGs) that do not present clinical manifestation [9, 10,11] .On the basis of this observation, nowadays the molecular diagnosis of SCA8 is only recommended for those families where the clinical affectation is well established.

Table 3 summarizes non-coding expansion SCAs.

FXTAS: Fragile X-Associated Tremor/Ataxia Syndrome

Fragile X syndrome (FXS) (OMIM #300624), which is the most common familial form of mental retardation (MR), is caused by the silencing of the *FMR1* gene, due to the presence of a full mutation (>200 CGG repeats) in the 5'UTR. The incidence of the syndrome is not known, but epidemiological studies indicate that it is responsible for MR in 1 in 4,000-6,000 males and in 1 in 7,000-10,000 females of European descendent.

In a study performed in Catalonia we found an incidence of 1:2,466 male and 1:8,333 females [12]. This difference between males and females is due to the reduced penetrance of FXS in females. It is also important to highlight the high incidence of premutation carriers (55-200 CGG repeats) in the general population, which has been estimated in 1 of 813 males and 1 of 259 females [13,14]

Individuals with the premutation are usually unaffected intellectually; however at least two different disorders, primary ovarian insufficiency (FXPOI), causing menopause before 40 years, and the fragile X-associated tremor/ataxia syndrome (FXTAS), have been described among them [8,15,16]. Both disorders do not share any clinical features with the FXS. Within the past few years, there has been a significant change in identifying and characterizing the *FMR1* premutation associated phenotypes. For example, the premutation has been associated with elevated *FMR1* mRNA levels and slight to moderate reductions in FMRP levels [17, 18].

FXTAS, was firstly identified among older male carriers of premutation alleles, and includes progressive action tremor and ataxia with associated radiological findings [8]. However, to date FXTAS has also been described among premutated women although it has been suggested that it occurs less frequently and that the phenotype is milder with older age at onset [16,19, 20]. An explanation for this difference is the presence of a second normal allele and a random X-inactivation of the premutated one; however, there may be additional sex-specific effects that reduce penetrance among females [8].

Table 2. Spinocerebellar ataxias: genetic classification
(CAG expansions in coding region)

	Chromosome	Normal range (CAG)	Premutated/ reduced penetrance	Expanded allele (CAG)	Protein	Pathophysiologica l mechanism	References
SCA1	6p22	6-39		40-83	Ataxin -1	Gain of function	[40]
SCA2	12q24.1	14-31	33-34	35-77	Ataxin-2	Gain of function	[41]
SCA3	14q32.1	12-40	41-53	54-86	Ataxin-3	Gain of function	[42]
SCA6	19p13	4-18		19-30	alpha-1A subunit calcium channel	Altered function	[43, 44]
SCA7	3p12-13	7-27	28-36	37-200	Ataxin-7	Gain of function	[45]
DRPLA	12p12	3-36	37-48	49-88	Atrophin 1	Gain of function	[46]
SCA17	6q27	25-43		44-63	TATA binding-protein	Altered function	[47]

Table 3. Spinocerebellar ataxias: genetic classification (expansions in non coding region)

	Chromosome	Nucleotide repeat	Normal range	Expanded allele	Protein	Pathophysiological mechanism	References
SCA8	13q21	CTG/CTA	16-92 CTG/CTA >250	107-250	Ataxin 8	Gain of function	[48]
SCA10	22q13	ATTCT	10-22	28-4500	Ataxin 10	Alterations in chromatin structure	[49,50]
SCA12	5q31-33	CAG	7-28	>65 CAG	Protein phosphatase 2	Gain of function	[51]
FXTAS	Xq27.3	CGG	<55	55-200	FMRP	Gain of function. Toxic RNA	[16]

Clinical symptoms of FXTAS syndrome appear in patients in their 50s or later, and they usually begin with an action tremor. After that, different findings including ataxia (balance problems with frequent falling), and more variably loss of sensation in the distal lower extremities and autonomic dysfunction (e.g., impotence, hypertension, and loss of bowel and bladder function) gradually progress. Cognitive deficits are also observed and include memory problems and executive function deficits with a gradual progression to dementia in some individuals [20]. Several of the patients have parkinsonian symptoms, including an intermittent resting tremor, masked facies, and increased tone. The patients also appear to experience psychological symptoms, including anxiety, reclusive behavior, and irritability or mood liability, all of which are likely to be related to the cognitive changes or to changes in the limbic area of the brain [8]. Magnetic resonance imaging (MRI) of patients affected with FXTAS demonstrates mild to moderate brain atrophy, which includes the cerebellum, with dilated ventricles and scattered periventricular white matter disease. In addition, hyperintensities of the middle cerebellar peduncles (MCP) on T2 is characteristic of FXTAS patients [8,21]. In a work performed by Jacquemont et al. [22] there are described three diagnostic categories in the diagnosis of FXTAS, each one reflecting combination of symptom types. The diagnostic categories are: i) "Definite" indicates the presence of one major radiological sign plus one major clinical symptom; ii) "Probable" indicates the presence of either one major radiology sign plus one minor clinical symptom or two major clinical symptoms; and iii) "Possible" indicates the presence of one minor radiological sign plus one major clinical symptom.

It has been estimated that at least one-third of all male carriers will develop a FXTAS syndrome, although there is significant variability in the progression of neurological dysfunction [8, 19, 22]. In our FXS families the estimated prevalence of FXTAS is ~40-50% for premutated male carriers and 16.5% for the permutated females [23].

The description and characterization of FXTAS syndrome is of great interest to the population, because the prevalence of *FMR1* premutation in the general population is relatively high. Several studies have been performed in order to determine the real role of FXTAS in undiagnosed adult patients with movement disorders. The results obtained in European populations ranges from 0% to 4% [24-27]. Our studies show an estimated FXTAS prevalence of 2% among patients presenting with ataxia of unknown etiology [28,29]. Although large studies are necessary to better define FXTAS prevalence in this kind of population, on the basis of premutation male frequency in general population, the prevalence of FXTAS has been estimated in ~1/3,000 males aged over 50 years of age (~1/10,000 males of all ages) [8].

FXTAS is an allelic disorder to the FXS, and therefore should be considered as a distinct neurodegenerative disorder. In fact, the molecular mechanism leading to FXTAS is distinct from the *FMR1* silencing mechanism and/or a deficit in FMRP operating in FXS. In premutated patients the *FMR1* gene is rarely silenced and FMRP levels are generally normal or only slightly lowered [8]. The only known molecular abnormality among premutation carriers is the presence of elevated levels of *FMR1* mRNA. The increased transcriptional activity of the *FMR1* gene seems to be positively correlated with the size of the CGG repeat. That is, CGG repeats in the upper range (100-200 CGG) result in average 5-fold elevation, whereas CGGs in the lower range (50-100 CGG) result in an average 2-fold elevation [17,18,30,31]. The presence of these elevated levels of abnormal (expanded CGG repeat) *FMR1* mRNA led to propose an RNA "toxic gain-of-function" model for FXTAS, in which the mRNA itself is causative of the neurological disorder [8,16,22,32]. The same mechanism has been proposed for myotonic dystrophy (DM1 and DM2), in which either the expanded repeat tract of CUG in DM1 or CCUG in DM2 sequestered CUG-binding proteins that disrupts mRNA processing of other genes or transport of other mRNAs [33]. In fact, this model has been demonstrated for DM1 by placing an expanded CTG tract in 3' UTR region of the *DMPK* mRNA in a transgenic mouse [34]. A part from this finding, FXTAS and myotonic dystrophy has another important similarity that will support the RNA gain-of-function mechanism for FXTAS syndrome. Both disorders show nuclear inclusions produced as a result of the binding proteins sequestered by the respectively mRNA, with a cytotoxic effect that lead to cell death. In a study performed by Greco and co-worker [32] eosinophilic intranucelear inclusions in neurons and astrocytes throughout the cortex and in deep cerebellar nuclei of FXTAS post-mortem samples were reported. Furthermore, in a subsequent study, there is described a highly significant association between CGG length and both the number of inclusions and the age of death, which correlates with the progressive character of the disease [35].

In order to test this RNA gain-of-function hypothesis for FXTAS, a mouse model has been generated in which the endogenous mouse CGG repeat was replaced by a human CGG tract carrying 98 CGGs [36]. Further studies of the brain of these expanded-repeat mice (at 20-72 weeks) evidenced elevated *Fmr1* mRNA levels and ubiquitin-positive intranuclear inclusions [37]. An increase was also observed in both the number and the size of the inclusions in specific brain region during the course of life [31]. The presence of inclusions in this mouse, that has normal levels of FMRP, provides evidences against a protein-deficiency model for FXTAS, and supports a direct role of the *Fmr1* gene, by either CGG expansion *per se* or by elevated *Fmr1* mRNA levels, in the pathology as the gain-of-function mechanism hypothesized.

To conclude, as the prevalence of premutated alleles is relatively high in general population, FXTAS may represent one of the more common monogenic causes of tremor, ataxia, and dementia. For this reason, it is probably that many carriers with FXTAS are being seen by a clinical specialist without awareness of the underlying genetic basis for the symptoms. The early diagnosis of those patients not only benefits themselves but also the rest of the family that should be advised for the FXS. Conventional mutations in SCA genes.

Table 4. Spinocerebellar ataxias: genetic classification (conventional mutations)

	Gen	Mutation	References
SCA5	*SPTBN2*	Missense, deletion	[52]
SCA11	*TTBK2*	Frameshift	[53]
SCA13	*KCNC3*	Missense	[54]
SCA14	*PRKCG*	Missense	[55]
SCA15/16	*ITPR1*	Missense, deletion	[56]
SCA20	---	Duplication	[57]
SCA27	*FGF14*	Missense, Frameship	[58]
SCA28	*AFG3L2*	Missense	[59,60]

Contrary to what we have seen in the previous two subtypes of SCAs (caused by a repeat expansion that is located inside or outside the coding region of the genes associated with the disease), in the present group of ADCAs the disease is attributed to a loss of function of the associated gene. Table 4 summarizes this subtype of SCAs.

We still ignore the real incidence of this subtype of SCAs because its molecular study is more complicated. The majority of the genes responsible for this kind of SCAs have been described by linkage studies in large families (ie. SCA11, SCA27, SCA28) [38], and their study is time and work consuming since a mutation screening of the whole gene is required.

Clinical traits of this subtype of SCAs defer of those caused by a triplet expansion. Anticipation and imprinting phenomena do not exist. The patients often had a childhood-onset, with a slowly disease progression that sometimes is associated with mental retardation. Furthermore, there is a loss of Purkinje cells and also a pure widespread cerebellar atrophy [38].

Genetic Diagnosis of SCAS

Nowadays the molecular diagnosis of SCAs is required in order to confirm a clinical suspicion. Molecular analysis allows rapid and accurate diagnosis and, at the same time, supports clinical suspicion and enables presymptomatic, prenatal and preimplantational diagnosis.

From the laboratory point of view, the study of SCAs caused by a trinucleotide expansion is easy since it is a recurrent mutation. When SCAs are caused by triplet expansions the analysis is focused in obtaining the accurate number of repeats. In these cases, molecular diagnosis is performed by Polymerase Chain Reaction (PCR), which allows the amplification of the repeat expansion, followed by the analysis of the fragment by electrophoresis on a capillary sequence. This strategy allows us to determine the exact repeat number (± 1 repeat), which is necessary in order to classify the status of the patient.

Clinical suspicion of a specific subtype of SCA, or at least classification in one of the four clinical groups, focuses the molecular diagnosis. Thus, careful and meticulous clinical assessment of patients is of significant help in choosing suitable molecular tests to ensure that the etiology is correctly defined. Routinely, the more prevalent types (SCA1, 2, 3, 6, 7, DRPLA and FXTAS) are studied; while SCA8, SCA12, and SCA17, are only studied if there is a strong clinical suspicion. As previously mentioned, the ethnic background of the patient is important since it is informative of which kind of SCA could be. Ataxias caused by conventional mutations are not studied unless there is a strong clinical suspicion or a linkage study in families with several affected members. In this situation, it is necessary to perform a direct mutational analysis of the responsible gene.

Genetic Counseling

The SCA patient should receive genetic counseling in person and it should be provided by an expert geneticist. Dynamic mutations are always inherited, and the affected individuals carry one allele in the mutation range (inherited from a carrier or affected parent) and the other in the normal one. Therefore, they have 50% of risk of transmitting the expanded allele to the offspring. Although there are some cases that appear to be sporadic cases, the possibility of a carrier parent with a reduced penetrance allele or with a premature death before disease onset can not completely ruled out.

Once the diagnosis in one individual is reached, familial study is always recommended. If possible, it is useful to enlarge the study to other family members in order to establish the origin (maternal or paternal) of the repeat expansion, since it might become a prognostic tool that might help the assessment of the severity and progression of the disease.

There are some homozygous (individuals carrying two mutated alleles) cases described; although clinically they are not much different than the heterozygous ones [9], in some reported cases of SCA3, the homozigosity enhances severity [39] and the genetic counseling is totally different since they have 100% risk of having an affected offspring.

Currently in ADCAs it is possible to offer a presymptomatic diagnosis with fully guaranties. In these cases the patient should be carefully advised by a genetic counselor, and in some cases the study must be done after a psychiatric examination and providing subsequent support. We have to keep in mind that SCAs are late-onset diseases with a progressive course for which, and at least up to date, there is not an effective treatment. For this reason one should be very cautious and conservative in performing childhood diagnosis, and therefore SCAs molecular diagnosis is not advisable when the patient is under the legal age.

Prenatal diagnosis is also accurate and reliable whenever the study of the affected parent is available. Nowadays the preimplantational diagnosis is also possible.

SCAs caused by conventional mutations are also inherited as a dominant trait, thus affected individuals have a 50% risk of transmitting the disease, although it exists the possibility of a sporadic case.

The only SCA inherited as an X-linked trait is FXTAS, and therefore the chances that a son or daughter will inherit the mutated chromosome from the mother are 50%, while all the daughters and none of the sons of carrier fathers receive the mutation. However, if the child will be affected or unaffected it depends on the dynamic mutation process, if the mutation remains in premutation range or progresses to the full mutation.

When a FXTAS female is detected, she is probably on her sixties, and therefore she has already reproduced. For this reason, genetic counseling is directed to her sons and daughters or grandchildren's since they have 50% risk of having a mentally retarded children. The most important in those cases is to strongly recommend familial studies.

On the other hand, when a FXTAS male is detected, the risk of expansion to the full mutation to their daughters is almost zero. Males with premutation pass on the premutation to all of their daughters, usually with only small expansions or contractions. All of his daughters will be normal carriers but with a high risk of having affected children in the next generation.

Molecular Analysis of SCAs

Reliability of the molecular study is very high (99%). The drawback is when the PCR only amplifies one allele in the normal range. This pattern can be attributed to a homozygosis individual with a two normal alleles or to an affected individual carrying a large expansion not amplified by PCR. One has to be very cautious and use all techniques and data available in order to discriminate between a homozygous individual and a non-amplified expanded allele due to technical difficulties.

For example in SCA7 large expansions are frequently seen, and 60% of individuals are homozygous for the allele with 10 CAGs repeats. In this situation, it is advisable to check the family pedigree and to repeat the technique including positive controls in order to discard a false negative.

According to Our Experience

From 2003 to 2010, 550 patients belonging to 541 independent family nucleuses have been studied for SCA1, 2, 3, 6, 7, 8 and DRPLA. FXTAS molecular test has been recently included into the SCAs study protocol based on the results published elsewhere [24,25,26,27] and in a study where we showed that 2% of the individuals with a movement disorder of unknown aetiology were carriers of a *FMR1* premuation. [28,29]

Among the 541 unrelated patients, 84 were positive (15.5%) for any of the 8 subtypes of ataxias studied. All of them have an index case with clinical diagnosis of cerebellar ataxia. If we consider those cases where a family history of ataxia was reported, 38 of 114 (33.3%) have a positive SCA molecular diagnosis. However, if we take into consideration sporadic, and cases without reported familial data, the positive rate is 10.8%. This fact exemplifies how important is to have a family history in order to perform a molecular study and how the probability of being affected decreases if no other family member has been previously diagnosed.

In our population the most prevalent subtype of ataxia is SCA3 (50.6% of the positive cases) followed by SCA2 (16.8%), SCA8 (12%), DRPLA and SCA1 with a 7.2%, FXTAS with a 1.95% and finally SCA6 and SCA7 with a 1.2%. Due to the geographical proximity to Portugal, it is expected SCA3 to be the most prevalent form of SCA in our country. Regarding SCA8 and DRPLA, it is remarkable their high prevalence in our population [9] when it has been mostly described in Asiatic countries.

Remarkably, two families presented expansions in homozygosis; one of them was a SCA1 and the other one a SCA8. In both cases, clinical traits were similar to the heterozygous cases.

Finally, we have performed 10 presymptomatic diagnoses in familial cases and 5 prenatal diagnoses. It is a low number of presymptomatic and prenatal cases if we compare to others pathologies such as Huntigton disease, where it is more accepted among affected families.

Conclusion

Here we review the different subtypes of SCAs and how they are clinically and genetically classified. Although the first approach is clinic, molecular diagnosis becomes essential, providing an accurate diagnostic and making possible genetic counselling, presymptomatic testing and prenatal diagnosis. It is important to highlight the fact that while a positive test establish a diagnostic, a negative one does not discard it. Furthermore it is important to understand the diversity of underlying mechanisms that give rise to the SCAs in order to identify possible therapeutic targets.

Acknowledgment

This work has received financial support from "Instituto Carlos III" (PS0900413).

References

[1]   Greenfield JG (1954) The Spino-cerebellar Degenerations. CC Thomas, Springfield, IL, pp 112.

[2]   Harding AE. Classifi cation of the hereditary ataxias and paraplegias. *Lancet* 1983; 1: 1151–55.

[3]   Harding AE. Hereditary spastic paraplegias. *Semin. Neurol.* 1993;13:333-6.

[4]   Teive HA, Munhoz RP, Raskin S, Werneck LC. Spinocerebellar ataxia type 6 in Brazil. *Arq. Neuropsiquiatr* 2008;66: 691–94.

[5]   Tarlac V, Storey E. Role of proteolysis in polyglutamine disorders. *J. Neurosci. Res.* 2003;74:406-16.

[6]   Matsuura T, Yamagata T, Burgess DL, Rasmussen A, Grewal RP, Watase K, Khajavi M, McCall AE, Davis CF, Zu L, Achari M, Pulst SM, Alonso E, Noebels JL, Nelson DL, Zoghbi HY, Ashizawa T. Large expansions of the ATTCT pentanucleotide repeat in spinocerebellar ataxia type 10. *Nat. Genet.* 2000;26:191-194.

[7]   Keren B, Jacquette A, Depienne C, Leite P, Durr A, Carpentier W, Benyahia B, Ponsot G, Soubrier F, Brice A, Héron D. Evidence against haploinsuffiency of human ataxin 10 as a cause of spinocerebellar ataxia type 10. *Neurogenetics* 2010; 11: 273–74.

[8]   Hagerman, PJ and Hagerman, RJ. The fragile-X premutation: a maturing perspective. *Am. J. Hum. Genet.* 2004;74:805-816.

[9]   Tazón B, Badenas C, Jiménez L, Muñoz E, Milà M. SCA8 in the Spanish population including one homozygous patient. *Clin. Genet.* 2002 Nov;62(5):404-9.

[10]   Duenas AM, Goold R, Giunti P. Molecular pathogenesis of spinocerebellar ataxias. *Brain* 2006;129:1357-1370.

[11]   Schols L, Bauer P, Schmidt T, Schulte T, Riess O. Autosomal dominant cerebellar ataxias: clinical features, genetics, and pathogenesis. *Lancet Neurol.* 2004; 3: 291–304.

[12]   Rife, M; Badenas, C; Mallolas, J; Jimenez, L; Cervera, R; Maya, A; Glover, G; Rivera, F; Mila, M. Incidence of fragile X in 5,000 consecutive newborn males. *Genet. Test* 2003;7:339-343.

[13]   Rousseau F, Rouillard P, Morel ML, Khandjian EW, Morgan K: Prevalence of carriers of premutation-size alleles of the FMRI gene–and implications for the population genetics of the fragile X syndrome. *Am. J. Hum. Genet.* 1995; 57: 1006– 1018.

[14]   Dombrowski C, Lévesque S, Morel ML, Rouillard P, Morgan K, Rousseau F. Premutation and intermediate-size FMR1 alleles in 10572 males from the general population: loss of an AGG interruption is a late event in the generation of fragile X syndrome alleles. *Hum. Mol. Genet.* 2002; 11: 371-378.

[15] Cronister A, Schreiner R, Wittenberger M, Amiri K, Harris K, Hagerman RJ: Heterozygous fragile X female: historical, physical, cognitive, and cytogenetic features. *Am. J. Med. Genet.* 1991; 38: 269– 274.

[16] Hagerman, RJ; Leehey, M; Heinrichs, W; Tassone, F; Wilson, R; Hills, J; Grigsby, J; Gage, B; Hagerman, PJ. Intention tremor, parkinsonism, and generalized brain atrophy in male carriers of fragile X. *Neurology* 2001;57:127-130.

[17] Tassone F, Hagerman RJ, Taylor AK, Gane LW, Godfrey TE, Hagerman PJ: Elevated levels of FMR1 mRNA in carrier males: a new mechanism of involvement in the fragile-X syndrome. *Am. J. Hum. Genet.* 2000a; 66: 6– 15.

[18] Tassone F, Hagerman RJ, Taylor AK, Mills JB, Harris SW, Gane LW, Hagerman PJ. Clinical involvement and protein expression in individuals with the FMR1 premutation. *Am. J. Med. Genet.* 2000b; 91: 144– 152.

[19] Jacquemont, S; Hagerman, RJ; Leehey, MA; Hall, DA; Levine, RA; Brunberg, JA; Zhang, L; Jardini, T; Gane, LW; Harris, SW; Herman, K; Grigsby, J; Greco, CM; Berry-Kravis, E; Tassone, F; Hagerman, PJ. Penetrance of the fragile X-associated tremor/ataxia syndrome in a premutation carrier population. *JAMA* 2004;291:460-469.

[20] Rodriguez-Revenga L, Pagonabarraga J, Gómez-Anson B, López-Mourelo O, Madrigal I, Xunclà M, Kulisevsky J, Milà M. Motor and mental dysfunction in mother-daughter transmitted FXTAS. *Neurology*. 2010 Oct 12;75(15):1370-6.

[21] Brunberg, JA; Jacquemont, S; Hagerman, RJ; Berry-Kravis, EM; Grigsby, J; Leehey, MA; Tassone, F; Brown, WT; Greco, CM; Hagerman, PJ. Fragile X premutation carriers: characteristic MR imaging findings of adult male patients with progressive cerebellar and cognitive dysfunction. *AJNR Am. J. Neuroradiol.* 2002;23:1757-1766.

[22] Jacquemont, S; Hagerman, RJ; Leehey, M; Grigsby, J; Zhang, L; Brunberg, JA; Greco, C; Des Portes, V; Jardini, T; Levine, R; Berry-Kravis, E; Brown, WT; Schaeffer, S; Kissel J; Tassone, F; Hagerman, PJ. Fragile X premutation tremor/ataxia syndrome: molecular, clinical, and neuroimaging correlates. *Am. J. Hum. Genet.* 2003;72:869-878.

[23] Rodriguez-Revenga L, Madrigal I, Pagonabarraga J, Xunclà M, Badenas C, Kulisevsky J, Gomez B, Milà M. Penetrance of FMR1 premutation associated pathologies in fragile X syndrome families. *Eur. J. Hum. Genet.* 2009 Oct;17(10):1359-62.

[24] Macpherson, J; Waghorn, A; Hammans, S; Jacobs, P. Observation of an excess of fragile-X premutations in a population of males referred with spinocerebellar ataxia. *Hum. Genet.* 2003;112:619-620.

[25] Zuhlke, Ch; Budnik, A; Gehlken, U; Dalski, A; Purmann, S; Naumann, M; Schmidt, M; Burk, K; Schwinger, E. FMR1 premutation as a rare cause of late onset ataxia-- evidence for FXTAS in female carriers. *J. Neurol.* 2004;251:1418-1419.

[26] Brussino, A; Gellera, C; Saluto, A; Mariotti, C; Arduino, C; Castellotti, B; Camerlingo, M; de Angelis, V; Orsi, L; Tosca, P; Migone, N; Taroni, F; Brusco, A. FMR1 gene premutation is a frequent genetic cause of late-onset sporadic cerebellar ataxia. *Neurology* 2005;64:145-147.

[27] Van Esch, H; Matthijs, G; Fryns, JP. Should we screen for FMR1 premutations in female subjects presenting with ataxia? *Ann. Neurol.* 2005;57:932-933.

[28] Rodriguez-Revenga L, Gómez-Anson B, Muñoz E, Jiménez D, Santos M, Tintoré M, Martín G, Brieva L, Milà M FXTAS in spanish patients with ataxia: support for female FMR1 premutation screening. *Mol. Neurobiol.* 2007 Jun;35(3):324-8.

[29]   Rodriguez-Revenga L, Santos MM, Sánchez A, Pujol M, Gómez-Anson B, Badenas C, Jiménez D, Madrigal I, Milà M. Screening for FXTAS in 95 Spanish patients negative for Huntington disease. *Genet. Test.* 2008 Mar;12(1):135-8.

[30]   Kenneson, A; Zhang, F; Hagedorn, CH; Warren, ST. Reduced FMRP and increased FMR1 transcription is proportionally associated with CGG repeat number in intermediate-length and premutation carriers. *Hum. Mol. Genet.* 2001;10:1449-1454.

[31]   Oostra, BA and Willemsen, R. A fragile balance: FMR1 expression levels. *Hum. Mol. Genet.* 2003;12:R249-R257.

[32]   Greco, CM; Hagerman, RJ; Tassone, F; Chudley, AE; Del Bigio, MR; Jacquemont, S; Leehey, M; Hagerman, PJ. Neuronal intranuclear inclusions in a new cerebellar tremor/ataxia syndrome among fragile X carriers. *Brain* 2002;125:1760-1771.

[33]   Mankodi, A and Thornton, CA. Myotonic syndromes. *Curr. Opin. Neurol.* 2002;15:545-552.

[34]   Mankodi, A; Logigian, E; Callahan, L; McClain, C; White, R; Henderson, D; Krym, M; Thornton, CA. Myotonic dystrophy in transgenic mice expressing an expanded CUG repeat. *Science* 2000;289:1769-1773.

[35]   Greco, CM; Berman, RF; Martin, RM; Tassone, F; Schwartz, PH; Chang, A; Trapp, BD; Iwahashi, C; Brunberg, J; Grigsby, J; Hessl, D; Becker, EJ; Papazian, J; Leehey, MA; Hagerman, RJ; Hagerman, PJ. Neuropathology of fragile X-associated tremor/ataxia syndrome (FXTAS). *Brain* 2006;129:243-255.

[36]   Bontekoe, CJ; Bakker, CE; Nieuwenhuizen, IM; van der Linde, H; Lans, H; de Lange, D; Hirst, MC; Oostra, BA. Instability of a (CGG)98 repeat in the Fmr1 promoter. *Hum. Mol. Genet..* 2001; 10: 1693-1639.

[37]   Willemsen R, Hoogeveen-Westerveld M, Reis S, Holstege J, Severijnen LA, Nieuwenhuizen IM, Schrier M, van Unen L, Tassone F, Hoogeveen AT, Hagerman PJ, Mientjes EJ, Oostra BA. The FMR1 CGG repeat mouse displays ubiquitin-positive intranuclear neuronal inclusions; implications for the cerebellar tremor/ataxia syndrome. *Hum. Mol. Genet.* 2003; 12: 949-959.

[38]   Durr A, Forlani S, Cazeneuve C, et al. Conventional mutations are associated with a diff erent phenotype than polyglutamine expansions in spinocerebellar ataxias. *Eur. J. Hum. Genet.* 2009; 17 (suppl 2): 335.

[39]   Carvalho DR, La Rocque-Ferreira A, Rizzo IM, Imamura EU, Speck-Martins CE.Homozygosity enhances severity in spinocerebellar ataxia type 3. *Pediatr. Neurol.* 2008 Apr;38(4):296-9.

[40]   Orr HT, Chung MY, Banfi S, Kwiatkowski TJ Jr, Servadio A, Beaudet AL, et al. Expansion of an unstable trinucleotide CAG repeat in spinocerebellar ataxia type 1. *Nature Genet.* 1993; 4: 221–6.41.

[41]   Wadia NH, Swami RK. A new form of heredo-familial spinocerebellar degeneration with slow eye movements (nine families). *Brain* 1971; 94:359-374.

[42]   Durr A, Stevanin G, Cancel G, et al. Spinocerebellar ataxia 3 and Machado-Joseph disease: clinical, molecular, and neuropathological features. *Ann. Neurol.* 1996; 39: 490–99.

[43]   Jodice C, Mantuano E, Veneziano L, et al. Episodic ataxia type 2 (EA2) and spinocerebellar ataxia type 6 (SCA6) due to CAG repeat expansion in the CACNA1A gene on chromosome 19p. *Hum. Mol. Genet.* 1997; 6: 1973–78.

[44] Stevanin G, Durr A, David G, et al. Clinical and molecular features of spinocerebellar ataxia type 6. *Neurology* 1997; 49: 1243–46.

[45] David G, Giunti P, Abbas N, et al. The gene forautosomal dominant cerebellara ataxia type II is located in a 5-cM region in 3p12-p13: genetic and physical mapping of the SCA7 locus. *Ann. J. Hum. Genet.* 1996; 59:1328-1336.

[46] Koide R, Ikeuchi T, Onodera O, Tanaka H, Igarashi S, Endo K, Takahashi H, Kondo R, Ishikawa A, Hayashi T, et al. Unstable expansion of CAG repeat in hereditary dentatorubral-pallidoluysian atrophy (DRPLA). *Nat. Genet.* 1994;6:9-13.

[47] Koide Koide R, Kobayashi S, Shimohata T, Ikeuchi T, Maruyama M, Saito M, Yamada M, Takahashi H, Tsuji S. A neurological disease caused by an expanded CAG trinucleotide repeat in the TATA-binding protein gene: a new polyglutamine disease? um *Mol. Genet.* 1999;8:2047-53.

[48] Koob MD, Moseley ML, Schut LJ, et al. An untranslated CTG expansion causes a novel form of spinocerebellar ataxia (SCA8). *Nat. Genet.* 1999; 21: 379–84.

[49] Zu L, Figueroa KP, Grewal R, Pulst S-M. Mapping of a new autosomal dominant spinocerebellar ataxia to chromosome 22. *Am. J. Hum. Genet.* 1999;64:594-599.

[50] Rasmunssen A, Matsuura T, Ruano L, et al. Clinical and genetic analysis of four Mexican families with spinocerebellar ataxia type 10. *Ann. Neurol.* 2001;50:234-239.

[51] Holmes SE, O'Hearn EE, McInnis MG, Gorelick-Feldman DA, Kleiderlein JJ, Callahan C, Kwak NG, Ingersoll-Ashworth RG, Sherr M, Sumner AJ, Sharp AH, Ananth U, Seltzer WK, Boss MA, Vieria-Saecker AM, Epplen JT, Riess O, Ross CA, Margolis RL. Expansion of a novel CAG trinucleotide repeat in the 5' region of PPP2R2B is associated with SCA12. *Nat. Genet.* 1999;23:391-2.

[52] Ikeda Y, Dick KA, Weatherspoon MR, et al Spectrin mutations cause spinocerebellar ataxia type 5. *Nat. Genet.* 2006;38:184–90.

[53] Houlden H,. Johnson J, Gardner-Thorpe C, Mutations in TTBK2, encoding a kinase implicated in tau phosphorylation, segregate with spinocerebellar ataxia type 11. *Nat. Genet.* 2007;39:1434–6.

[54] Waters MF, Minassian NA, Stevanin G, Figueroa KP, Bannister JP, Nolte D, Mock AF, Evidente VG, Fee DB, Müller U, Dürr A, Brice A, Papazian DM, Pulst SM. Mutations in voltage-gated potassium channel KCNC3 cause degenerative and developmental central nervous system phenotypes. *Nat. Genet.* 2006;38:447–51.

[55] Chen DH, Brkanac Z, Verlinde CL, Tan XJ, Bylenok L, Nochlin D, Matsushita M, Lipe H, Wolff J, Fernandez M, Cimino PJ, Bird TD, Raskind WH. Missense mutations in the regulatory domain of PKC gamma: a new mechanism for dominant nonepisodic cerebellar ataxia. *Am. J. Hum. Genet.* 2003;72:839–49.

[56] Hara K, Shiga A, Nozaki H, Mitsui J, Takahashi Y, Ishiguro H, Yomono H, Kurisaki H, Goto J, Ikeuchi T, Tsuji S, Nishizawa M, Onodera O. Total deletion and a missense mutation of ITPR1 in Japanese SCA15 families. *Neurology* 2008;71:547–51.

[57] Knight MA, Hernandez D, Diede SJ, Dauwerse HG, Rafferty I, van de Leemput J, Forrest SM, Gardner RJ, Storey E, van Ommen GJ, Tapscott SJ, Fischbeck KH, Singleton AB. A duplication at chromosome 11q12.2–11q12.3 is associated with spinocerebellar ataxia type 20. *Hum. Mol. Genet.* 2008;17:3847–53.

[58] van Swieten JC, Brusse E, de Graaf BM, Krieger E, van de Graaf R, de Koning I, Maat-Kievit A, Leegwater P, Dooijes D, Oostra BA, Heutink P. A mutation in the fibroblast

growth factor 14 gene is associated with autosomal dominant cerebellar ataxia [corrected]. *Am. J. Hum. Genet.* 2003;72:191–9.

[59] Mariotti C, Brusco A, Di Bella D, Cagnoli C, Seri M, Gellera C, Di Donato S, Taroni F. Spinocerebellar ataxia type 28: a novel autosomal dominant cerebellar ataxia characterized by slow progression and ophthalmoparesis. *Cerebellum.* 2008;7:184-8.

[60] Cagnoli C, Mariotti C, Taroni F, Seri M, Brussino A, Michielotto C, Grisoli M, Di Bella D, Migone N, Gellera C, Di Donato S, Brusco A. SCA28, a novel form of autosomal dominant cerebellar ataxia on chromosome 18p11.22-q11.2. *Brain.* 2006;129(Pt 1):235-42.

In: Genetic Diagnoses
Editor: R. J. Sarma, pp. 115-119

ISBN: 978-1-61324-866-9
© 2012 Nova Science Publishers, Inc.

Chapter V

Genetics of Left Ventricular Noncompaction

Radha J. Sarma [*]

Clinical Medicine, Los Angeles County USC Medical Center
Los Angeles, California, USA

Left ventricular noncompaction (LVNC) is still an unclassified cardiomyopathy according to the World Health Organization classification of cardiomyopathies, but in 2006 the American Heart Association classified this entity as a primary cardiomyopathy of genetic origin. LVNC is characterized by a distinctive spongy appearance due to excessive trabeculation of the myocardium and deep intertrabecular recesses that communicate with the left ventricular cavity. Elucidating a genetic basis for the phenotypic expression of LVNC has been difficult, since both familial and sporadic forms have been described. It appears that the pediatric presentation may have a distinct genetic basis as compared to the adult presentation. After the first case of isolated LVNC was reported 25years ago, much has been published about this entity. Even now, there is no consensus on the diagnostic criteria, and the diagnosis is based on the morphological features identified by cardiac imaging studies or at autopsy. Main complications seen in this condition are heart failure thromboembolism and arrhythmias. Association of neurological conditions has been reported with left ventricular hypertrabeculation. Different genes found to be associated with LVNC are (1) taffazin (TAZ), (2) alfa-dystrobrevin (DTNA), (3) Cypher/ZASP (LDB3) (4) lamin A/C (LMNA), (5) SCN5A, (6) MYH7 and (8) MYBPC3. There is also significant overlap in the phenotypes of the genetically mediated cardiomyopathies. Thus, LVNC can occur as LVNC with dilated or hypertrophic cardiomyopathy. Although further research is needed to elucidate the genetic basis of LVNC, at this time it seems clear that there is considerable genetic heterogeneity involved. In three of the largest series of patients, the rate of familial involvement was 18%, 25%, and 33%. In the familial cases, autosomal dominance is more common than sex-linked inheritance. While genetic testing is not routinely recommended at this time, the Heart Failure Society of America practice guidelines recommend clinical screening of all first degree

---

[*] Email: sarma@usc.edu, Phone: 626 590 8948 (cell), Phone: 323 409 8669 (office).

relatives of affected patients for LVNC. Genetic testing is to be considered for affected patients with a firm diagnosis of LVNC to help with family screening. Prognosis seems to be better in the recent publications likely due to improved therapeutic strategies. Standardization of diagnostic criteria is essential to continue further research and reporting results in LVNC. Genetic research still needs to be continued to further our knowledge of this genetic disorder.

Background: The first case of spongy myocardium was described by Engberding [1] in 1984. Because of the similarity of this abnormal myocardium, to the early embryonic heart, the name "left ventricular noncompaction" was proposed by Chin and Perloff [2] in 1990. The assumption that failure or arrest of the normal left ventricular compaction at the embryonic level is the reason for the isolated LVNC, was never confirmed and there are cases with acquired LVNC in adult life [3]. This cardiomyopathy is still considered unclassified by the WHO classification of cardiomyopathies [4]. In year 2006, the American heart Association task force classified LVNC as one of the genetically mediated cardiomyopathies [5]. In the past few years, there is a steady increase in the number of publications on the genetic association of LVNC [6-11]. However, only a small number of patients with LVNC have identifiable genetic mutations [12].

Genetics: Several investigators reported the familial occurrence of isolated LVNC in about 20-50% of adult cases [2, 13-16]. LVNC in the children appears to have x linked transmission [6, 17], while in adults it is autosomal dominant type of transmission [18]. There is considerable over lap of cases of LVNC with dilated cardiomyopathy and hypertrophic cardiomyopathy in the same family suggesting some common genetic mutations in the affected members [16-20].

At present, there is no consensus on the clinical criteria to diagnose LVNC [21-23]. The distinct morphology with the two layered appearance of the left ventricle, with a thin compacted epicardial layer and highly trabeculated thick endocardial layer, where the deep inter trabecular recesses communicate freely with the cavity is typical of LVNC, and the criteria published by Jenni et al (ref) appear to be used most popularly. Most common presenting symptoms are heart failure, arrhythmias or thromboembolic complications [2, 15-17, 24]. Since the condition is both familial [25-28] and sporadic [29, 30] and the morphology of the left ventricular myocardium is known to change over time , it is not clear, if this cardiomyopathy can be clinically identified with certainty. Hence, there is growing need for obtaining family history in at least three generations of the first-degree relatives and having clinical screening using the noninvasive imaging tests such as an echocardiogram or cardiac magnetic resonance imaging of the relatives [14, 22-23].

Many investigators published extensive details of the genetic mutations causing the different cardiomyopathies [12, 33]. It is clear from these studies that there is significant overlap in these mutations causing the different cardiomyopathies. At present, there is no specific mutation, which is definitive to identify the type of cardiomyopathy.

Until recently, genetic tests are not widely available for screening large number of patients and relatives. Because of the lack of specificity as well as availability, genetic testing is not routinely recommended, except when there are strong clinical correlates [34, 35]. In the past two years, there are many publications reporting the availability of these genetic tests, and the need for the clinicians to use genetic testing in the hypertrophic and dilated cardiomyopathies, including LVNC [14, 32, 33]. Electrocardiogram, echocardiography and tissue Doppler may help guide the type of genetic testing to be considered, because of the association of the particular morphology on echocardiogram, abnormalities on the electrocardiogram and the strong association with some select genes[32]. It seems that these tests may even help identify subjects before they develop full-fledged form of the disease. Although there is more research done on the hypertrophic cardiomyopathy, now it is increasingly obvious that the hypertrophic cardiomyopathy, dilated cardiomyopathy and LVNC all share some common genetic mutations [14, 32, 33].

Conclusion: It is about time that clinicians especially cardiologists, become familiar with the genetic abnormalities associated with the different primary genetic cardiomyopathies, and start using some of these genetic tests and use genetic councilors in their clinical practice. Genetic counseling started in the year 1969, but The National Society of Genetic Councilors (NSGC) came into existence ten years later [36]. They have expanded their role from prenatal counseling to pediatric and adult medicine over the years. Most major hospitals in United States and Canada do have these counselors available to help the cardiologists or other clinicians to help with their patients. In case, family members fear discrimination either socially or by insurance companies, they need to be informed of the Genetic information Nondiscrimination Act (GINA) that later became a law [32].

The future practice of genetic cardiomyopathies will be greatly influenced, by the genomic advances in medicine and the availability of commercial genetic tests. Education of cardiologists is equally important to understand the basics of genetically mediated cardiomyopathies and their clinical management including screening of the family members. Eventually genetic information may also affect the choice of drugs i.e., gene-specific therapy, as there is enough evidence to show that genetics do alter the drug responses [33].

References

[1]   Engberding R, Bender F: Identification of a rare congenital anomaly of the myocardium by two-dimensional echocardiography: persistence of isolated myocardial sinusoids. *Am. J. Cardiol.* 53:1733-1734, 1984.

[2]   Chin TK, Perloff JK, Williams RG, et al: Isolated noncompaction of left ventricular myocardium. A study of eight cases. *Circulation* 82:507-13, 1990.

[3]   Freedom R, Yoo S, Perrin D, et al: The morphological spectrum of ventricular noncompaction. *Cardiol. Young* 15:345-364, 2005.

[4]   Richardson P, McKenna W, Bristow M, Maisch B, Mautner B, O'Connell J, Olsen E,.Thiene G, Goodwin J, Gyarfas I, Martin I, Nordet P : Report of the 1995 World Health Organization/ International Society and Federation of Cardiology Task Force on the Definition and Classification of cardiomyopathies. *Circulation* 93(5):841-2, 1996.

[5]   Maron BJ, Towbin JA, Thiene G, et al: Contemporary definition and classification of the cardiomyopathies: an American Heart Association Scientific Statement from the Council on Clinical Cardiology, Heart Failure and Transplantation Committee; Quality of Care and Outcomes Research and Functional Genomics and Translational Biology Interdisciplinary Working Groups; and Council on Epidemiology and Prevention. *Circulation* 113:1807-16, 2006.

[6]   Bleyl SB, Mumford BR, Brown-Harrison MC, et al: Xq28-linked noncompaction of the ventricular myocardium: prenatal diagnosis and pathologic analysis of affected individuals. *Am. J. Med. Genet.* 72:257–265, 1997

[7]   Ichida F, Tsubata S, Bowles K, et al. Novel gene mutations in patients with left ventricular noncompaction or Barth syndrome. *Circulation* 2001;103:1256-1263.

[8]   Sasse-Klaassen S, Probst S, Gerull F, Oechslin E, et al. Novel gene locus for autosomal dominant left ventricular noncompaction maps to chromosome 11p15. *Circulation* 2004;109:2720-2723.

[9]   Budde BS, Binner P, Waldmuller S et al: Noncompaction of the Ventricular Myocardium Is Associated with a De Novo Mutation in the b-Myosin Heavy Chain Gene. PLoS ONE 2(12): e1362. doi:10.1371/journal.pone.0001362.

[10] Klaassen S, Probst S, Oechslin E et al: Mutations in Sarcomere Protein Genes in Left Ventricular Noncompaction. *Circulation.* 117:2893-2901, 2008.

[11] Dellefave L, McNally EM: Sarcomere Mutations in Cardiomyopathy, Noncompaction, and the Developing Heart. Editorial. *Circulation.* 117:2847-2849, 2008.

[12] Zaragoza M, Arbustini E, Narula J: Noncompaction of the left ventricle: primary cardiolyopathy with an elusive genetic etiology. *Curr. Opin. in Ped.* 19:619-627, 2007.

[13] Aras D, Tufekcioglu O, Erfun K, et al. Clinical features of isolated ventricular noncompaction in adults long-term clinical course, echocardiographic properties, and predictors of left ventricular failure. *J. Card Fail* 2006; 12: 726-733.

[14] Hoedemaekers YM, Caliskan K, Michels M, et al. The importance of genetic counseling, DNA diagnostics, and cardiologic family screening in left ventricular noncompaction cardiomyopathy. *Circ. Cardiovasc. Genet.* 3:232-9,2010.

[15] Lofiego C, Biagini E, Pasquale F, et al: Wide spectrum of presentation and variable outcomes of isolated left ventricular non-compaction. *Heart* 93:65-71, 2007.

[16] Murphy RT, Thaman R, Gimeno Blanes J, et al: Natural history and familial characteristics of isolated left ventricular non-compaction. *Eur. Heart J.* 24:187-92, 2005.

[17] Ichida F, Hamamichi Y, Miyawaki T, et al: Clinical features of isolated noncompaction of the ventricular myocardium: long-term clinical course, hemodynamic properties, and genetic background. *J. Am. Coll Cardiol.* 34:233-240, 1999.

[18] Sasse-Klaassen S, Gerull B, Oechslin E, et al. Isolated noncompaction of the left ventricular myocardium in the adult is an autosomal dominant disorder in the majority of patients. *Am. J. Med. Genet.*: 119A: 162-167. 2003.

[19] Biagini E, Ragni L, Ferlito M, et al: Different types of cardiomyopathy associated with isolated ventricular noncompaction. *Am. J. Cardiol..* 98: 821–824, 2006.

[20] Xing Y, Ichida F, Matsuoka T, et al: Genetic analysis in patients with left ventricular noncompaction and evidence for genetic heterogeneity. *Mol. Genet. Metab.* 88:71-77, 2006.

[21] Kohli S, Pantazis A, Shah J, et al. Diagnosis of left-ventricular non-compaction in patients with left-ventricular systolic dysfunction: time for a reappraisal of diagnostic criteria?. *Eur. Heart J.* 29:89–95, 2008.

[22] Engberding R, Yelbuz T, Breithardt G: Isolated noncompaction of the left ventricular myocardium. A review of the literature two decades after the initial case description. *Clin. Res. Cardiol.* 96:1–8, 2007.

[23] Oechslin E, Jenni R.: Left ventricular non-compaction revisited : a distinct phenotype with genetic heterogeneity? *Eur. Heart J.* doi:10.1093/eurheartj/ehq508, 2010.

[24] Oechslin E, Attenhofer Jost C, Rojas J, et al: Long-Term Follow-up of 34 Adults with Isolated Left Ventricular Noncompaction: A Distinct Cardiomyopathy with Poor Prognosis. *J. Am. Coll Cardiol.* 36: 493-500, 2000.

[25] Chung T, Yiannikas J,. Lee L C L, Lau, GT, and Kritharides L: Isolated Noncompaction Involving the Left Ventricular Apex in Adults. *Am. J. Cardiol.*; 94:1214–1216, 2004.

[26] Koh YY, Seo YU, Woo JJ, et al: Familial Isolated Noncompaction of the Ventricular Myocardium in Asymptomatic Phase. *Yonsei Med. J.,* 45:931-935, 2004

[27] Lorsheyd A, Cramer MJ, Velthuis BK, Vonken EJ, Van Der Smagt J, Van Tintelen P, Hauer RN. Familial occurrence of isolated non-compaction cardiomyopathy. *Eur. J. Heart Fail.* 8:826-31, 2006.

[28] Xia S, Wang H, Zhang X Zhu J and tang X: Clinical presentation and Genetic Analysis of a Five generation Chinese Family with Isolated Left Ventricular Noncompaction. *Inter. Med.;* 577-583, 2008.

[29] Ovadia M, Duque K S: Sporadic Isolated Left Ventricular Noncompaction. *Dread Disease or Not. PACE* 30: 455-457, 2007.

[30] Moura C, Hillion Y, Dalkha-Dahmane F, et al.: Isolated non-compaction of the myocardium diagnosed in the fetus: two sporadic and two familial cases. *Cardiol. Young*: 12, 278-283, 2002.

[31] Germans T, Dijkmans PA, Wilde AA, et al. Prominent crypt formation in the inferoseptum of a hypertrophic cardiomyopathy mutation carrier mimics noncompaction cardiomyopathy. *Circulation* 2007; 115:e610–e611.

[32] Bos M J., Towbin, J A, and Ackerman, M J: Diagnostic, prognostic, and therapeutic indication of genetic testing for hypertrophic cardiomyopathy. *JACC,* 54, 201–211, 2009.

[33] Callis TE, Jensen BC, Weck KE , Willis MS :Evolving molecular diagnostics for familial cardiomyopathies: at the heart of it all. *Expert Rev. Mol. Diagn.* 10, 329-351, 2010.

[34] Hershberger R, Lindenfeld J, Mestroni L, et al: Genetic Evaluation of Cardiomyopathy- A Heart Failure Society of America Practice Guideline. *J. Card Fail* 15:83-97, 2009.

[35] Robin N, Tabereaux P, Benza R, et al: Genetic Testing in Cardiovascular Disease. *J. Am. Coll Cardiol* 50:727-737, 2007.

[36] Skrzynia C, Demo EM, Baxter SM: Genetic Counseling and Testing for Hypertrophic Cardiomyopathy: An Adult Perspective. *J. of Cardiovasc. Trans. Res.* 2:493–499, 2009.

In: Genetic Diagnoses
Editor: R. J. Sarma, pp. 121-142

ISBN: 978-1-61324-866-9
© 2012 Nova Science Publishers, Inc.

Chapter VI

A Genetic Diagnosis to Type 1 Diabetes?

Jose Luis Santiago, Laura Espino-Paisán and Elena Urcelay
Hospital Universitario San Carlos, Madrid, Spain

Abstract

Type 1 Diabetes (T1D) is a complex trait caused by autoimmune destruction of islet beta cells in the pancreas resulting of the interaction between genetic and environmental factors. Despite enormous advances in the study of T1D there is still no cure and both genetic and environmental factors remain undefined. To date the therapy was mainly aimed at controlling hyperglycemia; however, recent therapeutic approaches include both the immunosuppression and immunetolerance for prevention and treatment of T1D. Autoreactive T cells have been proposed as target of this strategy because they are responsible for the autoimmune beta cell destruction. It is well known that preventing or avoiding the secondary effects of these therapies require a precise diagnosis of the autoimmune diseases. Hence, a genetic diagnosis would be suitable in order to prevent T1D and to achieve a correct immunotherapy.

The major T1D susceptibility genes, the HLA class II loci (HLA-DRB1 and HLA-DQB1) on chromosome 6p21 act in combination with other non-HLA genes across the genome. Recently, genome-wide association studies have identified over 40 chromosome regions outside the HLA as being associated with T1D. Some of these non-HLA loci have been implicated in other autoimmune diseases like celiac disease, rheumatoid arthritis or multiple sclerosis. The fact that distinct genetic variants are shared by different autoimmune diseases suggests that common immunological mechanisms are involved in the etiology of these autoimmune diseases.

The aim of this chapter is to analyze the T1D associated genes that have been replicated in different populations and whether they could lead to the genetic diagnosis. For this purpose we performed an analysis of three main established susceptibility genes in a Spanish cohort. We found that the cumulative presence of predisposition variants of the mentioned genes increases significantly the risk to disease.

Introduction

Type 1 diabetes mellitus (T1D) is a multifactorial autoimmune disorder resulting from selective destruction of the insulin producing beta cells in the pancreatic islets. This process is caused by the infiltration of the islets of Langerhans by dendritic cells, macrophages and auto-aggressive T-cells [1]. The risk of developing T1D is determined by a complex interaction between multiple genes and environmental factors, which both unfortunately remain undefined to date. The autoimmune process can be evidenced by the presence of autoantibodies to beta cell antigens, which appear almost in the 85% of patients who develop T1D and can be detected even 5 years before developing clinical manifestation of the disease [2]. These autoantibodies are merely markers of the destruction, because it is well known that beta-cell destruction is T-cell mediated. Apart from the classical disease-related autoantibodies that have been shown to predict clinical T1D: islet cells antibodies (ICA), anti-insulin, anti GAD (Glutamic Acid Descarboxylase) and anti islet-associated antigen 2 (IA2); recently a new autoantibody has been defined, anti ZnT8 (zinc cation efflux transporter) [3]. This transporter is specifically expressed in islet beta-cells, where it is associated with the insulin secretion pathway. The presence of multiple islet autoantibodies is highly predictive of future T1D, and it is associated with a progressive loss of insulin secretion during the preclinical stage [4].

The diagnosis of diabetes mellitus is based on clinical features; hyperglycemia is the main sign. However, the specific diagnosis of T1D is sometimes very difficult and most specific criteria would be useful to the diagnosis of this pathology. Moreover, by the time symptoms that lead to clinical diagnosis of T1D appears, most of the beta cells have been lost. Hence, prediction and prevention of T1D should be crucial targets with the aim to delay the onset of disease and ultimately reduce the future increases in diabetes incidence [5].

Genetic Predisposing Factors

The susceptibility to T1D is strongly associated with genes in the major histocompatibility complex (MHC), but the genetic basis of this complex trait is far to be known. Both linkage and association studies have identified some candidate genes that explain part of the familiar aggregation . Recently, high density genome wide association (GWA) studies have been one of the best tools that have contributed to improve our understanding of the genetic basis of autoimmune diseases such as T1D [6,7]. This genotyping analysis covers the majority of the common variants in the genome and establishes the relationship between markers across the human genome and predisposition to disease. Two essential findings have made possible the GWA studies. First, the international HapMap project [8] which provides patterns of genome variations and linkage disequilibrium (LD) in four different populations, facilitating both the design and analysis of association studies. Second, the availability of genotyping chips with hundred of thousands of single nucleotide polymorphisms (SNPs) making the study for thousand of cases and controls technically possible [9,10]. For T1D there has been a rapid expansion in the numbers of loci implicated in predisposition and the GWA studies have established to date more than 30 non-HLA regions [6,7,11]. Despite this high number of loci implicated in T1D susceptibility, most of them with modest effect, there are six genes-regions for which there is strong evidence of association with T1D. All of them have been previously robustly replicated by independent research groups:

The HLA class II genes on 6p21 in the Major Histocompatibility Complex (MHC).
The gene coding insulin (INS) on 11p15
The CTLA4 (cytotoxic T-lymphocite associated 4) gene located on 2q33
The PTPN22 (protein tyrosine phosphatase, non-receptor type 22) on 1q13
The interleukin-2 alpha chain (IL2RA) region on 10p15
The IFIH1/MDA5 (interferon-induced helicase 1) gene on 2q24

1. The MHC Class II Genes

The MHC is probably the most studied region in the human genome due to the involvement of this locus in autoimmune, infectious and inflammatory diseases, and transplantation. Although the MHC region is associated with almost every autoimmune disease, to date the causal variant remains unknown because it is very difficult to establish the precise contribution of the multiple genes and the multiple alleles for some of these genes as well as the extensive linkage disequilibrium (LD) that exists throughout this region.

There is no doubt that the major genetic susceptibility to T1D arises from the MHC [12-14]; in fact, close to 50% of the total genetic contribution to disease is attributable to this region [15,16]. The main genes involved in T1D susceptibility are located on MHC and they are the class II loci: HLA-DRB1, -DQA1 and DQB1. To date, most of the genetic studies in T1D have been carried out in white Caucasian populations and in all of them the haplotypes which consistently have demonstrated disease predisposition were DRB1*03-DQ2 y DRB1*04-DQ8. In fact, more than 90% of Caucasian subjects with T1D carry at least one of the two aforementioned haplotypes. However, several studies have reported that not all DRB1*03-DQ2 haplotypes predispose equally to disease [17-21]. Two conserved extended DRB1*03-DQ2 haplotypes or ancestral haplotypes (AH) were associated with diabetes susceptibility; the AH8.1 (also known as COX) and the AH18.2 (also known as QBL) [22]. Both haplotypes carry the same DRB1 and DQ alleles; however, the aforementioned studies have reported that AH18.2 haplotype confers significantly higher risk to T1D. This increased susceptibility must be due to presence of an additional gene on AH18.2 different from classical MHC class II. The characterization of this gene has been difficult mainly because the MHC region contains numerous genes with complex allelic and genetic structure, and also some of these genes are the most polymorphic of the genome, apart from the fact that this region shows high LD.

Furthermore, the MHC class II loci also confer the strongest protection from T1D. In Caucasian and Japanese population the protective haplotype is DRB1*1501-DQ6. Such protection dominates even in the presence of the high-risk susceptibility MHC class II alleles, although it is not absolute [23].

Despite the essence of the MHC association in T1D remains to be determined, it is clear that this region has to be considered as the first candidate in genetic diagnosis of T1D because these genes are the most important genetic factors in determining genetic risk or protection.

2. The Insulin Gene (INS)

Being T1D an autoimmune reaction to insulin-producing cells in the pancreas, and taking into account that the presence of anti-insulin antibodies in serum is one of the hallmarks of the disease, the insulin gene was early considered as a candidate risk factor and was, with HLA haplotypes, one of the most consistently replicated regions associated with T1D [24,25].

The insulin gene (INS) is located on chromosome 11, region 11p15.5. There are several polymorphisms in the gene, but the marker most associated with T1D contains a variable number of tandem repeats (VNTR) and is located 596 base pairs upstream of the INS locus. The consensus sequence of this VNTR is a group of 14 nucleotides and three classes of alleles are commonly established: class I or short class alleles have 26 to 63 repetitions, and class III alleles 140 to 210 repetitions. Class II includes the intermediate alleles with 64 to 139 repetitions. These alleles are less frequent and not clearly associated with disease [26].

Class III and class I alleles have been found associated to T1D in opposite ways. Class III alleles are protective, even in the presence of a susceptibility allele, and they cause a slight decrease of insulin mRNA expression in the pancreas but a strong increase of expression in the thymus. Class I alleles are associated to lower expression of insulin in the thymus and cause susceptibility to the disease [26]. These data point to the regulation of self-tolerance in the thymus as a milestone in the development of T1D pathology. By inducing higher levels of insulin expression in the thymus, class III alleles would allow a stronger negative selection and deletion of insulin-reactive T lymphocyte clones, preventing them to escape the thymus and begin the autoimmune reaction [26-28].

Reactivity to insulin alone is not enough to develop T1D, but it has been observed that individuals with the susceptibility polymorphism have a higher rate of insulin autoantibodies [29]. This condition, added to other genetic polymorphisms and as yet undetermined environmental factors, is most probably underlying the pathogenic mechanisms of the disease.

3. The CTLA4 (Cytotoxic T-Lymphocyte Associated 4) Gene

The cytotoxic T lymphocyte associated antigen-4 (CTLA-4) is a T cell surface receptor that binds the antigen-presenting cell membrane receptors CD80 and CD86, also known as B7.1 and B7.2. Antagonist of CTLA-4 is CD28, also expressed in the surface of T cells and also having CD80 and CD86 as ligands.

Binding of each molecule has opposite effects in the T cell. While CD28 is a costimulatory molecule and sends activating signals to the T cell, CTLA-4 binding to CD80 and CD86 activates a phosphatase cascade that inactivates the T cell, downregulating proliferation and cytokine production. Activation and suppression of T cell function are closely related processes. When a T cell receives activating signals through CD28, the same stimulus triggers the synthesis of CTLA-4, thus creating a negative feedback that prevents it from going into an uncontrolled activation process. On the other hand, T regulatory cells with an emergent role in suppression of autoimmunity express high levels of CTLA-4 that would contribute to the modulatory role of this subset of cells [30]. Interestingly, mutation or depletion of CTLA-4 leads to lymphoproliferative disorders in mice, remarking the importance of this membrane receptor in modulating the cell immune response [31,32].

For their roles controlling T cell activation, both CD28 and CTLA4 genes were interesting candidates for the study of genetic polymorphisms involved in autoimmune diseases like T1D.

CD28 and CTLA4 genes co-localize in chromosome 2 (2q33 region). Study of the LD in this region defines two blocks: one including the CTLA4 gene and the 5' region of ICOS (inducible co-stimulator) and other holding only the CD28 gene. The first studies found markers linked to T1D in the CTLA4 region [33]. Due to the strong LD, it could not be ascertained whether these signals came from markers in CTLA4 or in the 5' end of ICOS. Other study found that changes in ICOS polymorphisms did not affect the functionality of the encoded protein, while the studied polymorphisms in CTLA4 had functional effects in the CTLA-4 receptor that could explain its association with the disease. Thus, these data supported that association could be limited exclusively to the CTLA4 gene, discarding any role of ICOS and CD28 polymorphisms in the T1D etiology [34].

CTLA4 has attracted interest for many years and multiple studies have established association or linkage between this region and autoimmune diseases, particularly T1D [6,33-37]. The CTLA4 gene has four exons and three introns. Exon 1 codes for the leader peptide of the protein, exon 2 delivers the ligand-binding domain, exon 3 is the transmembrane domain and exon 4 the cytoplasmic tail. Two of the most studied and replicated polymorphisms in CTLA4 are rs231775 (+A49G), located in exon 1, and rs3087243 (C-318T, also known as CT60) in the promoter region.

The A allele of rs231775, a non-synonymous SNP located in exon 1 of the CTLA4 gene, codes for a threonine in position 17 of CTLA-4, forming a threonine-X-asparagin glycosilation site. The mutant G allele provides an alanine in the same position and breaks the glycosilation site, so it should be expected that this substitution had measurable consequences. In vitro experiments found that the G-allele derived protein is aberrantly glycosilated and ends up in lower levels of membrane-bound CTLA-4 [34].

The change CT60, a transition from a cytosine to a timine in position -318 of the CTLA4 gene promoter, is correlated to a higher activity of the promoter and, consequently, a higher expression of the CTLA-4 protein.

These two polymorphisms are the best studied, characterized and replicated in different populations, but there is still no proof that the association of this genetic region with T1D is attributable to one of these polymorphisms [35], nor any of them have been clearly identified as etiologic factors [34].

The CTLA4 gene can also undergo alternative splicing. Depending on the presence or absence of exon 3 in the resulting mRNA, two different isoforms are synthesized: one including the transmembrane domain and expressed in the membrane of T cells and a second soluble isoform commonly named sCTLA-4 that lacks the transmembrane domain. It is believed that the soluble isoform contributes to downregulate the activation of T cells by binding to CD80-CD86 and preventing the stimulation of CD28. Ueda et al found a correlation between high levels of sCTLA-4 in serum and the protective A allele in CT60 polymorphism [34]. By mechanisms as yet unknown, the protective allele augments the levels of sCTLA-4 mRNA and so patients with this genotype have higher levels of free sCTLA-4 in serum that contribute to control the activation of the immune system.

4. The PTPN22 Gene

The PTPN22 gene (protein tyrosine phosphatase, non-receptor type 22), located in chromosome 1 (region 1p13), encodes a lymphoid-specific phosphatase, LYP, which is an important downregulator of T cell activation. It is expressed mainly in T cells, but also in B cells, NK cells, macrophages, monocytes and dendritic cells and a PTPN22 polymorphism, C1858T, was found for the first time associated with T1D [38]. This polymorphisms is a non-synonymous SNP that causes a substitution from arginine to tryptophan in the 620 aminoacid of the encoded protein (R620W). The minor allele of this variant confers predisposition for T1D and was consistently replicated in independent populations [39-46]. The same allele was subsequently found associated with several autoimmune disorders, among them rheumatoid arthritis [47-51], systemic lupus eritematosus [49], Wegener´s granulomatosis [52] and myasthenia gravis [53]. The fact that a gene with a clear role in the modulation of T cell development and activation is associated with different autoimmune diseases seems to indicate that common biological pathways may be involved in the etiology of these pathologies.

Functional studies have revealed that LYP increases phosphatase activity when allele 1858T is present [54]. This gain of function mutant suppresses T cell signaling more efficiently and leads to a failure in apoptosis of autoreactive T cells and to an insufficient activity of regulatory T cells [55]. Since dysregulated autoaggresive T cells have been described as the main responsible for the beta cell destruction, the aforementioned polymorphism became an interesting candidate.

5. The Interleukin-2 Alpha Chain
Receptor Gene (IL2RA)

It has been long established that the imbalance between Th1 and Th2 cytokines plays a crucial role in the regulation of the immune response and also in the pathogenesis of autoimmune diseases [56-58]. Thus, the Th1 and Th2 cytokine genes and their receptors might be considered good candidates to modify the risk of these diseases. The main function of interleukin-2 (IL-2) is to promote proliferation and activation of both CD4 and CD8 T cells [59]. In fact, the IL-2 pathway has an essential role in the modulation of T cell regulation, and its function in suppressing the T cell immune response, probably involving regulatory T cells (Treg), has been suggested as a potential mechanism in the pathogenesis of autoimmune diseases [60]. It is known that IL-2 has an indispensable role in maintaining self-tolerance by supporting the growth, survival and function of Treg cells and the number of these cells are low in animal models which develop severe systemic inflammation with autoimmune components [61,62].

The IL2RA gene has been reported associated with T1D [6,63,64], but it is clear that it is interesting to study the other components of IL-2 pathway: the interleukin-2 gene (IL-2) on 4q27 and the IL-2 beta receptor gene (IL2RB) located on 22q13. Several polymorphisms located on the three aforementioned regions have been reported associated with T1D

[6,63,65] and other autoimmune diseases, such as celiac disease [66], rheumatoid arthritis (RA) [65] and multiple sclerosis (MS) [67].

To date, only the chromosomal region 10p15 showed convincing evidence for genetic association. However, we think that 4q27 and 22q13 must be considered as factors that deserve further study in T1D genetic predisposition.

6. The IFIH1/MDA5 Gene

Epidemiological studies have suggested the involvement of viral infection in T1D risk in genetically susceptible individuals [68-70]. The environmental factors would operate as a trigger in subjects with a background of genetic susceptibility and the high incidence of T1D in many countries over the past decades seems to indicate an increased environmental pressure on susceptibility genotypes.

The IFIH1/MDA5 gene (interferon-induced helicase 1) encodes the interferon beta inducible RNA helicase MDA-5, also known as helicard or IFIH1, which is implicated in the innate immune response to microbial pathogens. This protein participates in the apoptosis of virus infected cells by acting as dsRNA receptor formed in the picornavirus and enterovirus infections [71]. This cytoplasmic viral detector transmits a signal by a caspase recruitment domain and activates intracellular pathways leading to the induction of proinflammatory cytokines which finally will end up in the activation of adaptative immunity [71]. In addition, several studies have reported that picornavirus infections were associated with a higher risk to suffer T1D [72-74] and MS [75,76].

The IFIH1/MDA5 gene is located in the chromosomal region 2q24 and a polymorphism on exon 15, rs19990760 (A946T), was reportedly associated with T1D for the first time by Smyth et al [77], and then replicated in a genome wide study which validated the case control and familiar studies [7]. The minor allele of this SNP showed a protective effect for both T1D and MS in an independent population [78] in agreement to the one reported by Smyth et al. Therefore, the IFIH1/MDA5 gene could be a good candidate in order to consider the role of environmental factors in the development of the autoimmune process.

7. Other Genes

The genetic model assumed nowadays for T1D is that in which a small number of genes have a large effects and a large number of genes have small effects. A number of additional regions demonstrating some evidence of linkage or association with T1D have been identified during the last few years. Next, we will comment three of these genes that have shown evidence of disease association and have been validated in independent populations.

a. The CAPSL and IL7R Genes

In the already mentioned study that Smyth et al described the IFIH1/MDA5 gene association with T1D, they also reported the protective effect of a polymorphism located in the CAPSL (calcyphosine-like) gene [77]. This gene is located on the chromosomal region 5p13 and the same polymorphism was replicated in a genome wide study that found two new polymorphisms on the interleukin-7 receptor gene (IL7R), which is located in the same LD block that CAPSL gene [7]. However, in another genome-wide study carried out by the Welcome Trust Case Control Consortium (WTCCC) the analysis of diabetic population did not show association with 5p13 region. This fact could be explained because the most associated SNP of the CAPSL gene was not included in the aforementioned study. On the other hand, the polymorphisms on IL7R gene previously found associated also showed a protective effect although they did not exceed the threshold of significance for these pangenomic analyses.

The functional effect of the protein encoded by the CAPSL gene still remains unknown; however, the function of IL7R has been clearly established. It is expressed almost exclusively in lymphoid cells and is essential for development and maintenance of the immune system [79]. IL7R is a specific IL7 receptor and this cytokine is required for thymic maturation and proliferation of lymphocytes [80]. In fact, mutations in this gene in human patients cause a severe combined immunodeficiency (SCID) [81] with the major deficiencies in T cell development, whereas B and NK cells are relatively normal in numbers [82].

b. KIAA0350/CLEC16A

Two genome-wide studies recently performed [6,83] have pointed out to the region 16p13 in the long arm of chromosome 16 as a T1D-associated locus. This region contains a gene, termed at first KIAA0350, but lately known as CLEC16A (C-type lectin domain family 16 gene A) for its product holds a predicted C-type lectin domain. Little is known about the function of this protein but it is almost exclusively expressed in cells of the immune system, particularly in antigen-presenting cells and in NK cells.

C-type lectins are calcium-dependent polysaccharide binding proteins widely involved in several aspects of the immune response, from adhesion (selectins) to endocytic receptors or membrane-bound lymphocyte lectins, group to which CLEC16A probably belongs.

CLEC16A is a big gene of approximately 237 kbp included in a LD block with no other genes. This block is surrounded by CIITA (class II mayor histocompatibility complex transactivator) and SOCS1, a suppressor of cytokine signaling that inhibits the kinase activity of Jak in the Jak-STAT route. Both adjacent genes would be interesting functional candidates, but the studies performed up to date have not been able to relate the signals detected in 16p13 with polymorphisms in any of them, letting CLEC16A alone as the susceptibility gene for T1D in the region.

Functional studies would be required to explain how the detected polymorphisms influence the immune response. The first studies have not evidenced differences in mRNA expression levels between normal and mutant alleles of the implicated polymorphisms [83], in contrast to what has been observed, for example, in the CTLA4 +A49G polymorphism.

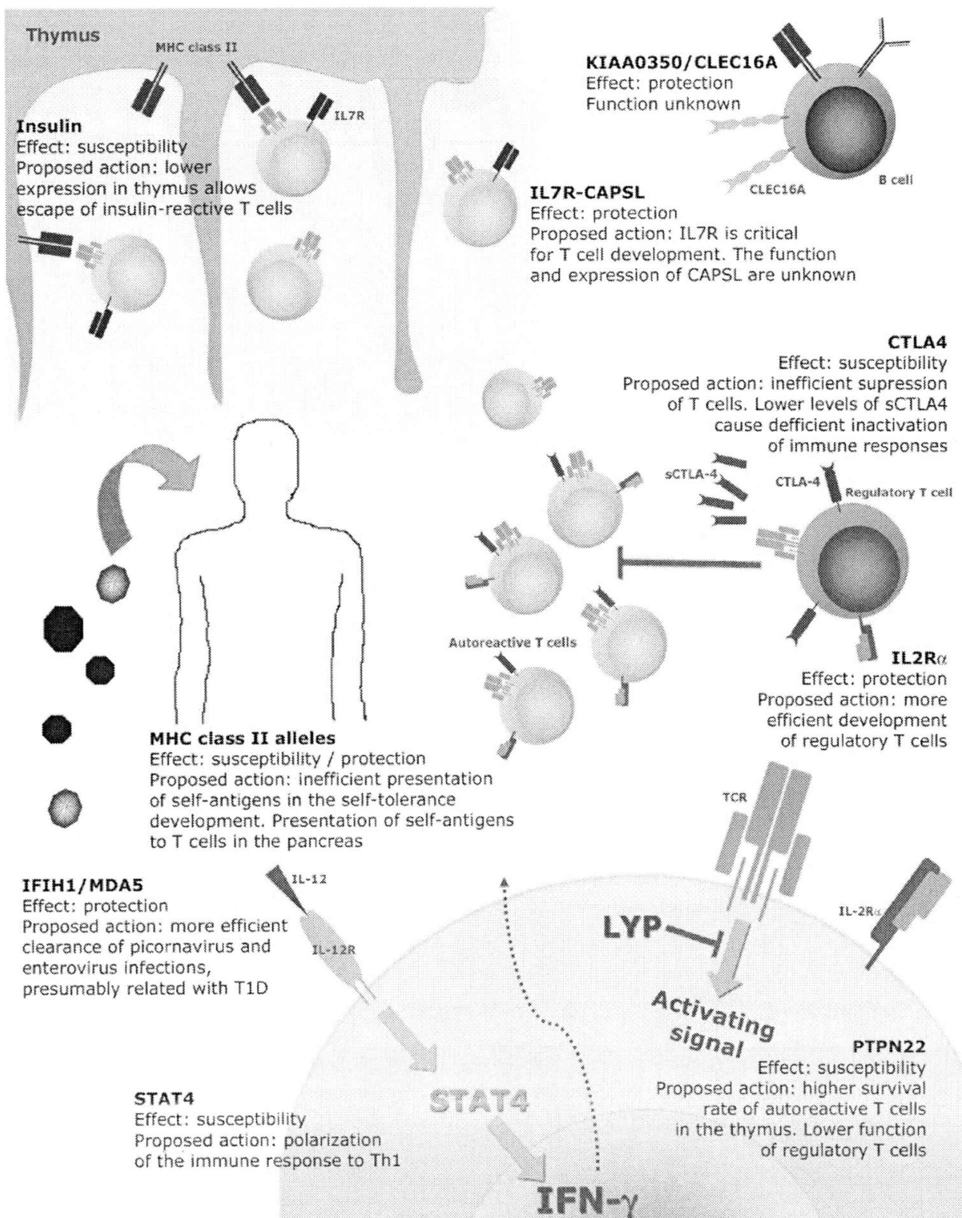

Figure 1. Proposed mechanisms for the main genes associated with T1D.

c. STAT4

STAT4 is a member of a group of cytoplasmic transcription factors called signal transducers and activators of transcription (STAT), proteins with a pivotal role in the transcription of activation signals.

Activating signals from membrane receptors can trigger enzymes termed JAK kinases and these molecules phosphorylate STATs. When phosphorylated, these transcription factors dimerize and travel to the nucleus, becoming active and starting the transcription of several genes. STAT4 participates in routes related to polarization of the immune response to Th1.

Throughout its action, signals from proinflammatory cytokines like IL-12, IL-15 or IL-23 are transmitted to T lymphocytes. The signaling route "proinflammatory cytokine $\rightarrow$ JAK $\rightarrow$ STAT4" finally activates the transcription of proinflammatory molecules such as interferon gamma. This cycle helps in the maintenance of the Th1 response.

Traditionally, the autoimmune response found in T1D patients has been considered a Th1 response and so, it is justifiable to search for candidate genes in the route of activation and signaling of the Th1 response, given that polymorphisms could overactivate the route and break the Th1/Th2 balance.

Several studies pointed out to polymorphisms in STAT4 as related to type 1 diabetes susceptibility [84-87]. Associations have also been found with several autoimmune diseases such as systemic lupus erithematosus, rheumatoid arthritis or Sjögren's disease [88,89]. Experiments in Stat4 null mice have been promising in uncovering the implication of this transcription factor in the mechanisms underlying autoimmune diseases. These mice have a lower rate of severe arthritis, hardly ever develop T1D and are resistant to experimental allergic encephalomyelitis, the mouse model for human multiple sclerosis [90]. Other experiments in non-obese diabetic (NOD) mice, the mouse model specific for T1D, showed that blocking of Stat4 prevented these mice from going into spontaneous diabetes [91].

STAT4 maps to chromosome 2, region 2q33, the same region as CTLA4, previously reviewed in this text and consistently associated with T1D and other autoimmune diseases like autoimmune thyroiditis. Thus, the region 2q33 constitutes a hot spot for T1D susceptibility.

Genetic Diagnosis in t1d

At this point and after the tremendous advances in the knowledge of the major susceptibility loci, we wonder whether the genetic diagnosis of T1D is a real possibility or just only illusion. The prediction of T1D has been one of the main targets since the first description of this autoimmune disease and it has been repeatedly suggested by researchers [92-94]. It is well known that the presence of specific risk alleles correlates with gradations in disease penetrance and subjects who carry the main susceptibility HLA haplotypes have a higher risk to develop T1D by the age of 15 years [95].

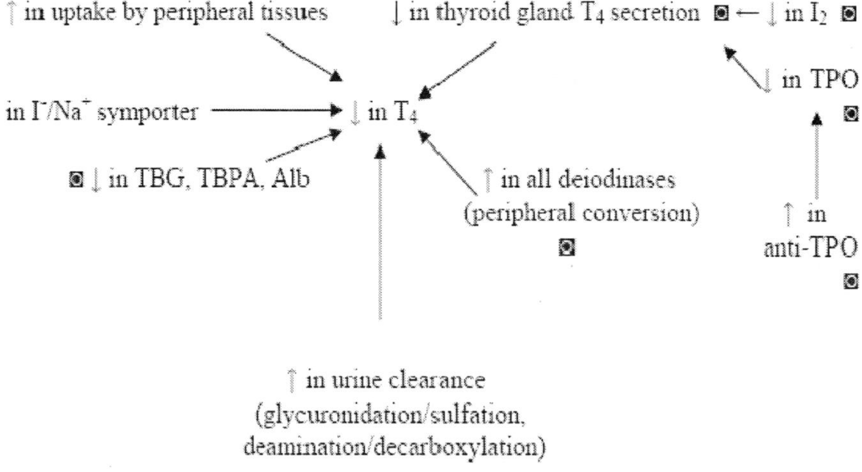

Figure 2. Relative risk for T1D measured as ORs in a Spanish cohort.

The large number of new risk variants for T1D that were identified from the GWA studies need to be fine mapped and their functional use well characterized for achieving both prediction and prevention of this autoimmune process.

In order to check the susceptibility conferred for some of the aforementioned genes we performed an analysis of the increasing risk to develop T1D depending on the alleles present in three susceptibility regions: HLA complex, the tyrosin phosphatase PTPN22 and the lymphocyte receptor CTLA4. This analysis was carried out in our study cohort that include 380 individuals diagnosed of T1D, insulin-dependent at the time of recruiting, with age at disease onset ranging from 1 to 59 years (median=16). Data were compared to a control cohort of 650 healthy subjects recruited from blood donors of the Madrid area. We studied the variation of risk to disease measured as odds ratio. For this purpose, we defined a reference group lacking of susceptibility alleles in the studied genes, and consecutive groups that carried susceptibility in one gene but did not in the rest, two genes, three genes and so forth (Figure 2). As expected, HLA is the region with a strongest contribution to disease susceptibility. The contributions of PTPN22 and the two polymorphisms in CTLA4 are similar, increasing the risk conferred by the HLA alleles. The presence of susceptibility alleles in the four polymorphisms analyzed increases the risk almost double than the HLA alleles by themselves.

In summary, the screening for susceptibility genes in the general population may play an efficient role in primary T1D prevention. However, we must be extremely precise in the identification of subjects under study because there are some aspects of the genetic method that we have to take into consideration. Next we will try to clarify why sometimes the genetic identification of the risk groups is difficult.

Genetic Variability: The Asian Case

A helpful diagnostic test to predict disease risk based on genetic variants must be carefully designed with one precaution in mind: the human genetic heterogeneity.

Recent genome wide studies with large populations are almost exclusively carried out in Caucasian subjects and so, those studies are detecting susceptibility variants only in Caucasian populations. Results must be carefully replicated in the remaining ethnic groups in order to detect different effects of a known variant, absence of the susceptibility variant in study or differences in the frequency of the susceptibility allele between populations.

A good example of this genetic heterogeneity would be the Asian population. Large studies have been conducted mainly in Japanese and Korean subjects and have unveiled the subtle differences between these populations and Caucasians. To illustrate this point, we will analyze some of the genes already mentioned and the results found by researchers in Asian subjects, mainly of Japanese and Korean descent.

HLA susceptibility haplotypes in Asians are an appealing example of human genetic heterogeneity. Frequencies of HLA haplotypes are variable among human populations, sometimes so dramatically different that alleles with high frequency in one population may be completely absent in another. Such is the case of T1D associated alleles. The susceptibility haplotypes in Caucasian subjects (DRB1*0301-DQB*0201 and DRB1*0401-DQB*0302), with frequencies around 5-7%, are very rare in Asian populations (<1%). However, frequencies of the protective haplotype (DRB1*1501-DQB*0602) are similar in both ethnic backgrounds (7-10%) [96].

Considering that the previously described susceptibility haplotypes account for 50% of the genetic load of the disease and that those haplotypes are practically inexistent in Asian population, it could be assumed that the frequency of the disease is much lower in Asians than in Caucasians. It is lower indeed (about ten times less frequent), but nevertheless abundant, and careful study of HLA haplotypes revealed new susceptibility alleles in Asiatic subjects, the relatively frequent DRB1*0405-DQB*0401 and DRB1*0901-DQB*0303, that in Caucasian populations are extremely rare and therefore would have passed undetected in Caucasian-based studies. Also, higher and lower susceptibility goes in different directions than in Caucasian haplotypes. Higher susceptibility is conferred by the DR9 haplotype in homozygous state, although DR9 in heterozygous state has no effect. On the other hand, the DR4 haplotype confers susceptibility in either homozygous or heterozygous dotation [97].

The insulin gene (INS) poses another interesting difference between Asian and Caucasian populations. Susceptibility is conferred by class I alleles of the VNTR near the INS promoter, and protection by longer class III alleles. In Caucasians, the frequency of class I and class III alleles is approximately 50% for both. This leads to think that having the abundant class I allele is an important factor to add to other susceptibility genes. However, in Asian populations the frequency of the class I susceptibility alleles is more than 95%, and so the overall population holds susceptibility to T1D in this locus. Through metanalysis, it has been concluded that the protective and susceptibility effects of both types of alleles are the same in both populations [98], but the diagnostic value in Asian populations would be much lower given that virtually everybody holds susceptibility to disease in this precise locus. The same happens in African populations, where the class III protective allele is absent.

Similar to the INS gene is the case of the PTPN22 gene. The best characterized variation, C1858T, is absent in large collections of Japanese and Korean subjects, which leads to think that the presence of the C1858T polymorphism in Asian populations is anecdotic, if it occurs at all. However, the same researchers found another polymorphism, located in the PTPN22 promoter (-1123GC) that, once analyzed, was found to be associated with the disease in the same subjects previously studied for the C1858T substitution. The promoter polymorphism is also present in Caucasian subjects, and association studies have pointed to -1123GC as another susceptibility polymorphism in Caucasian population [99]. PTPN22 would be the case of a gene associated in two populations (Caucasians and Asians) through two different polymorphisms, the "Caucasian" C1858T and the "shared" -1123GC. An opposite case is that of the Small Ubiquitin-related Modifier or SUMO4, not reviewed in the present text but named here for its populational differences. The most studied variant, which causes a substitution between metionine and valine in aminoacid 55 (M55V), has been consistently associated with T1D and other autoimmune diseases in Asian subjects but failed to show association in European populations [100]. Thus, this gene would be interesting in the analysis of the disease risk in Asian subjects, but it is not useful when it comes to predict risk for a Caucasian patient. A genetic test to calculate risk to T1D would have to take into account these and other ethnic variations to achieve success. It is necessary to precisely characterize the genetic basis of the disease in different ethnic groups and the effects of the susceptibility variants in order to achieve prediction and apply the best treatment, should this be preventive or posterior to disease onset. A test based in the presence of DR3/DR4 haplotypes, insulin class I alleles and the genotyping of the C1858T polymorphism in PTPN22 gene would be useless for a Japanese patient, but probably very informative for an European one.

Therapeutic Strategies in T1D

The risk of T1D complications is mainly determined by the total lifetime glucose levels, therefore the treatment of this disease is focused on reaching and maintaining normoglycemia. Unfortunately, the insulin therapy does not assure normal glucose blood levels for many patients, which contributes to increase the risk of heart disease or kidney failure. On the other hand, the cardiovascular morbidity continues high among diabetic patients with close monitoring of glucose levels [101]. All these facts have forced researchers to raise new effective strategies both in treatment and prevention of T1D in order to develop the best therapy for all patients.

1. Treatments

The drugs that classically have been used in the treatment of T1D have not been totally effective. In addition, the insulin replacement therapy is not curative and most of these patients develop complications for years. In fact, diabetes has become a major cause of premature illness and death in Western populations, mainly due to the increased risk of cardiovascular disease. Besides, among chronic diseases, diabetes has a major impact on healthcare costs, which is expected to raise over time due to the increasing prevalence of T1D [102,103].

Today, the only way to permanently restore the normoglycemia without hypoglycemia is to provide patients with additional beta cells. This can be achieved by transplanting an intact pancreas or by islets transplantation. However, the surgical complications in the whole pancreas transplantation and the shortage of functional beta cells from available donors in the islet transplantation are the major limiting factors in these therapeutic approaches. Therefore, new methods that preserve or even promote regeneration of beta cells mass are really needed.

During the last few years significant progress has been made in stem cell knowledge and their use in therapy has become one of the most promising cures for some dramatic diseases. In T1D several research groups using embryonic stem cells [104,105], stem cells from bone marrow [106] and mesenchymal stem cells [107] have reported that these cells are capable to differentiate towards insulin producing cells in vitro. However, the therapy based on stem cells for the treatment of T1D is still in the beginning and new techniques are needed for identification of appropriate stem cells and adequate assays that allow to assess their differentiation in vivo.

2. Prevention and Genetics

The current therapy for T1D and the associated complications of this treatment have an enormous impact on public health care resources, apart from the economic cost. During the last few years novel preventive and regenerative therapies have been the ultimate goal of new treatments to preserve or even promote regeneration of the patient's remaining beta cell mass and to delay the age at T1D onset.

To date, T1D is an autoimmune disease for which no prevention therapy has been used routinely. Different clinical trials are currently under development and despite some of them are at present in phase II, most of these therapeutic interventions are animal studies. These novel approaches for the prevention of autoimmune diabetes have involved both antigen specific therapies [108], systemic immunosuppression [109], antigen non specific therapies [110,111] and combinations of different drugs[112].

Genetics can help in the development of new preventive therapies and even achieve a greater success by individualizing the elected treatment. Here we comment two examples of ongoing clinical trials: one aimed to use insulin as a sensitizing therapy and a second focused on the development of a treatment with recombinant sCTLA4 as a suppressor of the immune system.

Insulin has been extensively studied as a primary autoantigen and as a susceptibility gene, but also as a potential preventive therapy. Many clinical trials are aiming at using the knowledge around insulin to develop an effective and non-aggressive preventive treatment. One of the main clinical trials in this field is the Diabetes Prevention Trial (DPT-1) in which oral and parenteral insulin are being tested in relatives of T1D patients with a predicted high risk of developing the disease.

The aim is to induce tolerance to insulin through a low-dose exogenous administration. Interestingly, results were promising in the group of patients that had high titers of insulin autoantibodies and were treated with oral insulin. A significant decrease in T1D incidence was observed in this group [113].

Production of insulin autoantibodies is theoretically triggered by genetic and environmental factors. Genetic factors could be partially explained by the effect of class I alleles in the VNTR polymorphism within the INS locus, associated with higher insulin autoantibody titers.

These data suggest that though T1D has one curative treatment (correcting the deficiency with exogenous insulin), preventive treatments must determine the exact nature of the immune response evaluating the susceptibility genes and autoantibodies involved.

Individualize the treatment would lead to stop or retard the specific immune response that is destroying the β-cell mass in each patient. If the findings of the DPT-1 prove to be consistent, then patients with INS class I alleles and high antibody titers would be good candidates to oral insulin therapy, while patients with class III alleles and medium-to-low antibody titers could spare the inconvenience of following a treatment that would probably make no difference to the final onset of diabetes.

Modulating therapies that would allow to control the autoimmune response is another focus. Through the characterization of the effects of CTLA4 polymorphisms on the encoded protein, it came to light that high levels of sCTLA4, the alternative splicing form of the receptor membrane, were associated with protection to T1D and other autoimmune diseases.

This protection was mediated through the modulating effect exerted by competitive binding of sCTLA4 to CD80-CD86 molecules, avoiding CD28 prostimulating molecules to be activated. Following this idea, clinical trials are in development aiming to use the properties of sCTLA-4 to control the activation of the immune system.

A fusion protein containing the CD80-CD86 binding motif of CTLA-4 is being tested in newly diagnosed T1D patients, hoping that the immunosuppressive properties of the molecule contribute to a better preservation of the remaining pool of functioning insulin-producing β-cells in the pancreas [28].

The main problem in the identification of suitable preventive strategies is the accurate definition of the risk groups. Despite the dramatic progress in understanding the genetic factors involved in T1D, the familial aggregation is low since close to 90% of patients do not have a familiar history of diabetes [114]. In addition, the mode of inheritance is not clear [93]. However, there is no doubt that genetic information can be extremely useful as a predictive tool in order to improve stratification of patients according to risk and to select the appropriate therapeutic targets.

Conclusion

Genetic information is a useful predictive tool to improve the correct stratification of patients according to risk and to establish an adequate therapy to prevent or delay the disease onset. Despite a better understanding of the potential genes implicated in T1D, enormous work remains to be developed to generate a genetic risk formula that takes into account both susceptibility and protection alleles implicated in T1D. Thus, as the familiar-cancer genetic counseling can suggest patients to take some preventive measures, in T1D this approach should be the first target in the future. Time has come to consider T1D as an ordinary autoimmune disease and treatment has to start before clinical symptoms appear. Once the genetic factors are known, the balance risk/benefit of preventive treatment will not create any doubt.

References

[1] Achenbach P, Bonifacio E, Koczwara K, Ziegler AG. Natural history of type 1 diabetes. Diabetes. 2005 Dec;54 Suppl 2:S25-31.

[2] Ziegler AG, Hummel M, Schenker M, Bonifacio E. Autoantibody appearance and risk for development of childhood diabetes in offspring of parents with type 1 diabetes: the 2-year analysis of the German BABYDIAB Study. Diabetes. 1999 Mar;48(3):460-8.

[3] Wenzlau JM, Juhl K, Yu L, Moua O, Sarkar SA, Gottlieb P, et al. The cation efflux transporter ZnT8 (Slc30A8) is a major autoantigen in human type 1 diabetes. Proc Natl Acad Sci U S A. 2007 Oct 23;104(43):17040-5.

[4] Verge CF, Gianani R, Kawasaki E, Yu L, Pietropaolo M, Jackson RA, et al. Prediction of type I diabetes in first-degree relatives using a combination of insulin, GAD, and ICA512bdc/IA-2 autoantibodies. Diabetes. 1996 Jul;45(7):926-33.

[5] von Boehmer H. Type 1 diabetes: focus on prevention. Nat Med. 2004 Aug;10(8):783-4.

[6] Genome-wide association study of 14,000 cases of seven common diseases and 3,000 shared controls. Nature. 2007 Jun 7;447(7145):661-78.

[7] Todd JA, Walker NM, Cooper JD, Smyth DJ, Downes K, Plagnol V, et al. Robust associations of four new chromosome regions from genome-wide analyses of type 1 diabetes. Nat Genet. 2007 Jul;39(7):857-64.

[8] A haplotype map of the human genome. Nature. 2005 Oct 27;437(7063):1299-320.

[9] Hirschhorn JN, Daly MJ. Genome-wide association studies for common diseases and complex traits. Nat Rev Genet. 2005 Feb;6(2):95-108.

[10] Wang WY, Barratt BJ, Clayton DG, Todd JA. Genome-wide association studies: theoretical and practical concerns. Nat Rev Genet. 2005 Feb;6(2):109-18.

[11] Cooper JD, Smyth DJ, Smiles AM, Plagnol V, Walker NM, Allen JE, et al. Meta-analysis of genome-wide association study data identifies additional type 1 diabetes risk loci. Nat Genet. 2008 Dec;40(12):1399-401.

[12] Nerup J, Platz P, Andersen OO, Christy M, Lyngsoe J, Poulsen JE, et al. HL-A antigens and diabetes mellitus. Lancet. 1974 Oct 12;2(7885):864-6.

[13] Noble JA, Valdes AM, Cook M, Klitz W, Thomson G, Erlich HA. The role of HLA class II genes in insulin-dependent diabetes mellitus: molecular analysis of 180 Caucasian, multiplex families. Am J Hum Genet. 1996 Nov;59(5):1134-48.

[14] Todd JA. Genetic control of autoimmunity in type 1 diabetes. Immunol Today. 1990 Apr;11(4):122-9.

[15] Davies JL, Kawaguchi Y, Bennett ST, Copeman JB, Cordell HJ, Pritchard LE, et al. A genome-wide search for human type 1 diabetes susceptibility genes. Nature. 1994 Sep 8;371(6493):130-6.

[16] Todd JA. Genetic analysis of type 1 diabetes using whole genome approaches. Proc Natl Acad Sci U S A. 1995 Sep 12;92(19):8560-5.

[17] Johansson S, Lie BA, Todd JA, Pociot F, Nerup J, Cambon-Thomsen A, et al. Evidence of at least two type 1 diabetes susceptibility genes in the HLA complex distinct from HLA-DQB1, -DQA1 and -DRB1. Genes Immun. 2003 Jan;4(1):46-53.

[18] Lie BA, Todd JA, Pociot F, Nerup J, Akselsen HE, Joner G, et al. The predisposition to type 1 diabetes linked to the human leukocyte antigen complex includes at least one non-class II gene. Am J Hum Genet. 1999 Mar;64(3):793-800.

[19] Nejentsev S, Gombos Z, Laine AP, Veijola R, Knip M, Simell O, et al. Non-class II HLA gene associated with type 1 diabetes maps to the 240-kb region near HLA-B. Diabetes. 2000 Dec;49(12):2217-21.

[20] Urcelay E, Santiago JL, de la Calle H, Martinez A, Mendez J, Ibarra JM, et al. Type 1 diabetes in the Spanish population: additional factors to class II HLA-DR3 and -DR4. BMC Genomics. 2005;6(1):56.

[21] Zavattari P, Lampis R, Motzo C, Loddo M, Mulargia A, Whalen M, et al. Conditional linkage disequilibrium analysis of a complex disease superlocus, IDDM1 in the HLA region, reveals the presence of independent modifying gene effects influencing the type 1 diabetes risk encoded by the major HLA-DQB1, -DRB1 disease loci. Hum Mol Genet. 2001 Apr 1;10(8):881-9.

[22] Degli-Esposti MA, Abraham LJ, McCann V, Spies T, Christiansen FT, Dawkins RL. Ancestral haplotypes reveal the role of the central MHC in the immunogenetics of IDDM. Immunogenetics. 1992;36(6):345-56.

[23. Ikegami H, Fujisawa T, Kawabata Y, Noso S, Ogihara T. Genetics of type 1 diabetes: similarities and differences between Asian and Caucasian populations. Ann N Y Acad Sci. 2006 Oct;1079:51-9.

[24] Bell GI, Horita S, Karam JH. A polymorphic locus near the human insulin gene is associated with insulin-dependent diabetes mellitus. Diabetes. 1984 Feb;33(2):176-83.

[25] Julier C, Hyer RN, Davies J, Merlin F, Soularue P, Briant L, et al. Insulin-IGF2 region on chromosome 11p encodes a gene implicated in HLA-DR4-dependent diabetes susceptibility. Nature. 1991 Nov 14;354(6349):155-9.

[26] Vafiadis P, Bennett ST, Todd JA, Nadeau J, Grabs R, Goodyer CG, et al. Insulin expression in human thymus is modulated by INS VNTR alleles at the IDDM2 locus. Nat Genet. 1997 Mar;15(3):289-92.

[27] Alizadeh BZ, Koeleman BP. Genetic polymorphisms in susceptibility to Type 1 Diabetes. Clin Chim Acta. 2008 Jan;387(1-2):9-17.

[28]Ounissi-Benkalha H, Polychronakos C. The molecular genetics of type 1 diabetes: new genes and emerging mechanisms. Trends Mol Med. 2008 Jun;14(6):268-75.

[29]Hermann R, Laine AP, Veijola R, Vahlberg T, Simell S, Lahde J, et al. The effect of HLA class II, insulin and CTLA4 gene regions on the development of humoral beta cell autoimmunity. Diabetologia. 2005 Sep;48(9):1766-75.

[30]Fife BT, Bluestone JA. Control of peripheral T-cell tolerance and autoimmunity via the CTLA-4 and PD-1 pathways. Immunol Rev. 2008 Aug;224:166-82.

[31]Tivol EA, Borriello F, Schweitzer AN, Lynch WP, Bluestone JA, Sharpe AH. Loss of CTLA-4 leads to massive lymphoproliferation and fatal multiorgan tissue destruction, revealing a critical negative regulatory role of CTLA-4. Immunity. 1995 Nov;3(5):541-7.

[32]Waterhouse P, Penninger JM, Timms E, Wakeham A, Shahinian A, Lee KP, et al. Lymphoproliferative disorders with early lethality in mice deficient in Ctla-4. Science. 1995 Nov 10;270(5238):985-8.

[33]Nistico L, Buzzetti R, Pritchard LE, Van der Auwera B, Giovannini C, Bosi E, et al. The CTLA-4 gene region of chromosome 2q33 is linked to, and associated with, type 1 diabetes. Belgian Diabetes Registry. Hum Mol Genet. 1996 Jul;5(7):1075-80.

[34]Ueda H, Howson JM, Esposito L, Heward J, Snook H, Chamberlain G, et al. Association of the T-cell regulatory gene CTLA4 with susceptibility to autoimmune disease. Nature. 2003 May 29;423(6939):506-11.

[35]Anjos SM, Tessier MC, Polychronakos C. Association of the cytotoxic T lymphocyte-associated antigen 4 gene with type 1 diabetes: evidence for independent effects of two polymorphisms on the same haplotype block. J Clin Endocrinol Metab. 2004 Dec;89(12):6257-65.

[36]Concannon P, Erlich HA, Julier C, Morahan G, Nerup J, Pociot F, et al. Type 1 diabetes: evidence for susceptibility loci from four genome-wide linkage scans in 1,435 multiplex families. Diabetes. 2005 Oct;54(10):2995-3001.

[37]Lee YJ, Lo FS, Shu SG, Wang CH, Huang CY, Liu HF, et al. The promoter region of the CTLA4 gene is associated with type 1 diabetes mellitus. J Pediatr Endocrinol Metab. 2001 Apr;14(4):383-8.

[38]Bottini N, Musumeci L, Alonso A, Rahmouni S, Nika K, Rostamkhani M, et al. A functional variant of lymphoid tyrosine phosphatase is associated with type I diabetes. Nat Genet. 2004;36(4):337-8. Epub 2004 Mar 7.

[39]Gomez LM, Anaya JM, Gonzalez CI, Pineda-Tamayo R, Otero W, Arango A, et al. PTPN22 C1858T polymorphism in Colombian patients with autoimmune diseases. Genes Immun. 2005;6(7):628-31.

[40]Ladner MB, Bottini N, Valdes AM, Noble JA. Association of the single nucleotide polymorphism C1858T of the PTPN22 gene with type 1 diabetes. Hum Immunol. 2005;66(1):60-4.

[41]Onengut-Gumuscu S, Ewens KG, Spielman RS, Concannon P. A functional polymorphism (1858C/T) in the PTPN22 gene is linked and associated with type I diabetes in multiplex families. Genes Immun. 2004;5(8):678-80.

[42]Qu H, Tessier MC, Hudson TJ, Polychronakos C. Confirmation of the association of the R620W polymorphism in the protein tyrosine phosphatase PTPN22 with type 1 diabetes in a family based study. J Med Genet. 2005;42(3):266-70.

[43]Santiago JL, Martinez A, de la Calle H, Fernandez-Arquero M, Figueredo MA, de la Concha EG, et al. Susceptibility to type 1 diabetes conferred by the PTPN22 C1858T polymorphism in the Spanish population. BMC Med Genet. 2007;8:54.

[44.Smyth D, Cooper JD, Collins JE, Heward JM, Franklyn JA, Howson JM, et al. Replication of an association between the lymphoid tyrosine phosphatase locus (LYP/PTPN22) with type 1 diabetes, and evidence for its role as a general autoimmunity locus. Diabetes. 2004;53(11):3020-3.

[45]Zheng W, She JX. Genetic association between a lymphoid tyrosine phosphatase (PTPN22) and type 1 diabetes. Diabetes. 2005;54(3):906-8.

[46]Zhernakova A, Eerligh P, Wijmenga C, Barrera P, Roep BO, Koeleman BP. Differential association of the PTPN22 coding variant with autoimmune diseases in a Dutch population. Genes Immun. 2005;6(6):459-61.

[47]Begovich AB, Carlton VE, Honigberg LA, Schrodi SJ, Chokkalingam AP, Alexander HC, et al. A missense single-nucleotide polymorphism in a gene encoding a protein tyrosine phosphatase (PTPN22) is associated with rheumatoid arthritis. Am J Hum Genet. 2004;75(2):330-7. Epub 2004 Jun 18.

[48]Carlton VE, Hu X, Chokkalingam AP, Schrodi SJ, Brandon R, Alexander HC, et al. PTPN22 genetic variation: evidence for multiple variants associated with rheumatoid arthritis. Am J Hum Genet. 2005;77(4):567-81. Epub 2005 Aug 10.

[49]Orozco G, Sanchez E, Gonzalez-Gay MA, Lopez-Nevot MA, Torres B, Caliz R, et al. Association of a functional single-nucleotide polymorphism of PTPN22, encoding lymphoid protein phosphatase, with rheumatoid arthritis and systemic lupus erythematosus. Arthritis Rheum. 2005;52(1):219-24.

[50]Simkins HM, Merriman ME, Highton J, Chapman PT, O'Donnell JL, Jones PB, et al. Association of the PTPN22 locus with rheumatoid arthritis in a New Zealand Caucasian cohort. Arthritis Rheum. 2005;52(7):2222-5.

[51]van Oene M, Wintle RF, Liu X, Yazdanpanah M, Gu X, Newman B, et al. Association of the lymphoid tyrosine phosphatase R620W variant with rheumatoid arthritis, but not Crohn's disease, in Canadian populations. Arthritis Rheum. 2005;52(7):1993-8.

[52]Jagiello P, Aries P, Arning L, Wagenleiter SE, Csernok E, Hellmich B, et al. The PTPN22 620W allele is a risk factor for Wegener's granulomatosis. Arthritis Rheum. 2005;52(12):4039-43.

[53]Vandiedonck C, Capdevielle C, Giraud M, Krumeich S, Jais JP, Eymard B, et al. Association of the PTPN22*R620W polymorphism with autoimmune myasthenia gravis. Ann Neurol. 2006;59(2):404-7.

[54]Vang T, Congia M, Macis MD, Musumeci L, Orru V, Zavattari P, et al. Autoimmune-associated lymphoid tyrosine phosphatase is a gain-of-function variant. Nat Genet. 2005 Dec;37(12):1317-9.

[55]Bottini N, Vang T, Cucca F, Mustelin T. Role of PTPN22 in type 1 diabetes and other autoimmune diseases. Semin Immunol. 2006 Aug;18(4):207-13.

[56]Cantagrel A, Navaux F, Loubet-Lescoulie P, Nourhashemi F, Enault G, Abbal M, et al. Interleukin-1beta, interleukin-1 receptor antagonist, interleukin-4, and interleukin-10 gene polymorphisms: relationship to occurrence and severity of rheumatoid arthritis. Arthritis Rheum. 1999 Jun;42(6):1093-100.

[57]Rabinovitch A. Immunoregulatory and cytokine imbalances in the pathogenesis of IDDM. Therapeutic intervention by immunostimulation? Diabetes. 1994 May;43(5):613-21.

[58]Sartor RB. Cytokines in intestinal inflammation: pathophysiological and clinical considerations. Gastroenterology. 1994 Feb;106(2):533-9.

[59]Gaffen SL, Liu KD. Overview of interleukin-2 function, production and clinical applications. Cytokine. 2004 Nov 7;28(3):109-23.

[60]Malek TR, Bayer AL. Tolerance, not immunity, crucially depends on IL-2. Nat Rev Immunol. 2004 Sep;4(9):665-74.

[61]Setoguchi R, Hori S, Takahashi T, Sakaguchi S. Homeostatic maintenance of natural Foxp3(+) CD25(+) CD4(+) regulatory T cells by interleukin (IL)-2 and induction of autoimmune disease by IL-2 neutralization. J Exp Med. 2005 Mar 7;201(5):723-35.

[62]Thornton AM, Donovan EE, Piccirillo CA, Shevach EM. Cutting edge: IL-2 is critically required for the in vitro activation of CD4+CD25+ T cell suppressor function. J Immunol. 2004 Jun 1;172(11):6519-23.

[63]Lowe CE, Cooper JD, Brusko T, Walker NM, Smyth DJ, Bailey R, et al. Large-scale genetic fine mapping and genotype-phenotype associations implicate polymorphism in the IL2RA region in type 1 diabetes. Nat Genet. 2007 Sep;39(9):1074-82.

[64]Vella A, Cooper JD, Lowe CE, Walker N, Nutland S, Widmer B, et al. Localization of a type 1 diabetes locus in the IL2RA/CD25 region by use of tag single-nucleotide polymorphisms. Am J Hum Genet. 2005 May;76(5):773-9.

[65]Zhernakova A, Alizadeh BZ, Bevova M, van Leeuwen MA, Coenen MJ, Franke B, et al. Novel association in chromosome 4q27 region with rheumatoid arthritis and confirmation of type 1 diabetes point to a general risk locus for autoimmune diseases. Am J Hum Genet. 2007 Dec;81(6):1284-8.

[66]van Heel DA, Franke L, Hunt KA, Gwilliam R, Zhernakova A, Inouye M, et al. A genome-wide association study for celiac disease identifies risk variants in the region harboring IL2 and IL21. Nat Genet. 2007 Jul;39(7):827-9.

[67]Maier LM, Lowe CE, Cooper J, Downes K, Anderson DE, Severson C, et al. IL2RA genetic heterogeneity in multiple sclerosis and type 1 diabetes susceptibility and soluble interleukin-2 receptor production. PLoS Genet. 2009 Jan;5(1):e1000322.

[68]Elfaitouri A, Berg AK, Frisk G, Yin H, Tuvemo T, Blomberg J. Recent enterovirus infection in type 1 diabetes: evidence with a novel IgM method. J Med Virol. 2007 Dec;79(12):1861-7.

[69]Lonnrot M, Korpela K, Knip M, Ilonen J, Simell O, Korhonen S, et al. Enterovirus infection as a risk factor for beta-cell autoimmunity in a prospectively observed birth cohort: the Finnish Diabetes Prediction and Prevention Study. Diabetes. 2000 Aug;49(8):1314-8.

[70]Salminen K, Sadeharju K, Lonnrot M, Vahasalo P, Kupila A, Korhonen S, et al. Enterovirus infections are associated with the induction of beta-cell autoimmunity in a prospective birth cohort study. J Med Virol. 2003 Jan;69(1):91-8.

[71]Kato H, Takeuchi O, Sato S, Yoneyama M, Yamamoto M, Matsui K, et al. Differential roles of MDA5 and RIG-I helicases in the recognition of RNA viruses. Nature. 2006 May 4;441(7089):101-5.

[72]Benoist C, Mathis D. Autoimmunity provoked by infection: how good is the case for T cell epitope mimicry? Nat Immunol. 2001 Sep;2(9):797-801.

[73]Skarsvik S, Puranen J, Honkanen J, Roivainen M, Ilonen J, Holmberg H, et al. Decreased in vitro type 1 immune response against coxsackie virus B4 in children with type 1 diabetes. Diabetes. 2006 Apr;55(4):996-1003.

[74]Yoon JW, Jun HS. Viruses cause type 1 diabetes in animals. Ann N Y Acad Sci. 2006 Oct;1079:138-46.

[75]Kriesel JD, White A, Hayden FG, Spruance SL, Petajan J. Multiple sclerosis attacks are associated with picornavirus infections. Mult Scler. 2004 Apr;10(2):145-8.

[76]Olson JK, Ercolini AM, Miller SD. A virus-induced molecular mimicry model of multiple sclerosis. Curr Top Microbiol Immunol. 2005;296:39-53.

[77]Smyth DJ, Cooper JD, Bailey R, Field S, Burren O, Smink LJ, et al. A genome-wide association study of nonsynonymous SNPs identifies a type 1 diabetes locus in the interferon-induced helicase (IFIH1) region. Nat Genet. 2006 Jun;38(6):617-9.

[78]Martinez A, Santiago JL, Cenit MC, de Las Heras V, de la Calle H, Fernandez-Arquero M, et al. IFIH1-GCA-KCNH7 locus: influence on multiple sclerosis risk. Eur J Hum Genet. 2008 Jul;16(7):861-4.

[79]Kondo M, Takeshita T, Higuchi M, Nakamura M, Sudo T, Nishikawa S, et al. Functional participation of the IL-2 receptor gamma chain in IL-7 receptor complexes. Science. 1994 Mar 11;263(5152):1453-4.

[80]Watanabe M, Ueno Y, Yajima T, Iwao Y, Tsuchiya M, Ishikawa H, et al. Interleukin 7 is produced by human intestinal epithelial cells and regulates the proliferation of intestinal mucosal lymphocytes. J Clin Invest. 1995 Jun;95(6):2945-53.

[81]Roifman CM, Zhang J, Chitayat D, Sharfe N. A partial deficiency of interleukin-7R alpha is sufficient to abrogate T-cell development and cause severe combined immunodeficiency. Blood. 2000 Oct 15;96(8):2803-7.

[82]Giliani S, Mori L, de Saint Basile G, Le Deist F, Rodriguez-Perez C, Forino C, et al. Interleukin-7 receptor alpha (IL-7Ralpha) deficiency: cellular and molecular bases. Analysis of clinical, immunological, and molecular features in 16 novel patients. Immunol Rev. 2005 Feb;203:110-26.

[83]Hakonarson H, Grant SF, Bradfield JP, Marchand L, Kim CE, Glessner JT, et al. A genome-wide association study identifies KIAA0350 as a type 1 diabetes gene. Nature. 2007 Aug 2;448(7153):591-4.

[84]Fung EY, Smyth DJ, Howson JM, Cooper JD, Walker NM, Stevens H, et al. Analysis of 17 autoimmune disease-associated variants in type 1 diabetes identifies 6q23/TNFAIP3 as a susceptibility locus. Genes Immun. 2009 Mar;10(2):188-91.

[85]Lee HS, Park H, Yang S, Kim D, Park Y. STAT4 polymorphism is associated with early-onset type 1 diabetes, but not with late-onset type 1 diabetes. Ann N Y Acad Sci. 2008 Dec;1150:93-8.

[86]Martinez A, Varade J, Marquez A, Cenit MC, Espino L, Perdigones N, et al. Association of the STAT4 gene with increased susceptibility for some immune-mediated diseases. Arthritis Rheum. 2008 Sep;58(9):2598-602.

[87]Zervou MI, Mamoulakis D, Panierakis C, Boumpas DT, Goulielmos GN. STAT4: a risk factor for type 1 diabetes? Hum Immunol. 2008 Oct;69(10):647-50.

[88]Korman BD, Alba MI, Le JM, Alevizos I, Smith JA, Nikolov NP, et al. Variant form of STAT4 is associated with primary Sjogren's syndrome. Genes Immun. 2008 Apr;9(3):267-70.

[89]Remmers EF, Plenge RM, Lee AT, Graham RR, Hom G, Behrens TW, et al. STAT4 and the risk of rheumatoid arthritis and systemic lupus erythematosus. N Engl J Med. 2007 Sep 6;357(10):977-86.

[90]Boyton RJ, Davies S, Marden C, Fantino C, Reynolds C, Portugal K, et al. Stat4-null non-obese diabetic mice: protection from diabetes and experimental allergic encephalomyelitis, but with concomitant epitope spread. Int Immunol. 2005 Sep;17(9):1157-65.

[91]Yang Z, Chen M, Ellett JD, Fialkow LB, Carter JD, McDuffie M, et al. Autoimmune diabetes is blocked in Stat4-deficient mice. J Autoimmun. 2004 May;22(3):191-200.

[92]Barker JM, Barriga KJ, Yu L, Miao D, Erlich HA, Norris JM, et al. Prediction of autoantibody positivity and progression to type 1 diabetes: Diabetes Autoimmunity Study in the Young (DAISY). J Clin Endocrinol Metab. 2004 Aug;89(8):3896-902.

[93]Concannon P, Rich SS, Nepom GT. Genetics of type 1A diabetes. N Engl J Med. 2009 Apr 16;360(16):1646-54.

[94]Eisenbarth GS. Prediction of type 1 diabetes: the natural history of the prediabetic period. Adv Exp Med Biol. 2004;552:268-90.

[95]Aly TA, Ide A, Humphrey K, Barker JM, Steck A, Erlich HA, et al. Genetic prediction of autoimmunity: initial oligogenic prediction of anti-islet autoimmunity amongst DR3/DR4-DQ8 relatives of patients with type 1A diabetes. J Autoimmun. 2005;25 Suppl:40-5.

[96]Ikegami H, Kawabata Y, Noso S, Fujisawa T, Ogihara T. Genetics of type 1 diabetes in Asian and Caucasian populations. Diabetes Res Clin Pract. 2007 Sep;77 Suppl 1:S116-21.

[97]Kawabata Y, Ikegami H, Kawaguchi Y, Fujisawa T, Shintani M, Ono M, et al. Asian-specific HLA haplotypes reveal heterogeneity of the contribution of HLA-DR and -DQ haplotypes to susceptibility to type 1 diabetes. Diabetes. 2002 Feb;51(2):545-51.

[98]Kawaguchi Y, Ikegami H, Shen GQ, Nakagawa Y, Fujisawa T, Hamada Y, et al. Insulin gene region contributes to genetic susceptibility to, but may not to low incidence of, insulin-dependent diabetes mellitus in Japanese. Biochem Biophys Res Commun. 1997 Apr 7;233(1):283-7.

[99]Kawasaki E, Awata T, Ikegami H, Kobayashi T, Maruyama T, Nakanishi K, et al. Systematic search for single nucleotide polymorphisms in a lymphoid tyrosine phosphatase gene (PTPN22): association between a promoter polymorphism and type 1 diabetes in Asian populations. Am J Med Genet A. 2006 Mar 15;140(6):586-93.

[100] Noso S, Ikegami H, Fujisawa T, Kawabata Y, Asano K, Hiromine Y, et al. Genetic heterogeneity in association of the SUMO4 M55V variant with susceptibility to type 1 diabetes. Diabetes. 2005 Dec;54(12):3582-6.

[101] Reichard P, Nilsson BY, Rosenqvist U. The effect of long-term intensified insulin treatment on the development of microvascular complications of diabetes mellitus. N Engl J Med. 1993 Jul 29;329(5):304-9.

[102] Hogan P, Dall T, Nikolov P. Economic costs of diabetes in the US in 2002. Diabetes Care. 2003 Mar;26(3):917-32.

[103] Zhang P, Engelgau MM, Norris SL, Gregg EW, Narayan KM. Application of economic analysis to diabetes and diabetes care. Ann Intern Med. 2004 Jun 1;140(11):972-7.

[104] Assady S, Maor G, Amit M, Itskovitz-Eldor J, Skorecki KL, Tzukerman M. Insulin production by human embryonic stem cells. Diabetes. 2001 Aug;50(8):1691-7.

[105] Lumelsky N, Blondel O, Laeng P, Velasco I, Ravin R, McKay R. Differentiation of embryonic stem cells to insulin-secreting structures similar to pancreatic islets. Science. 2001 May 18;292(5520):1389-94.

[106] Poulsom R, Alison MR, Forbes SJ, Wright NA. Adult stem cell plasticity. J Pathol. 2002 Jul;197(4):441-56.

[107] Tang DQ, Cao LZ, Burkhardt BR, Xia CQ, Litherland SA, Atkinson MA, et al. In vivo and in vitro characterization of insulin-producing cells obtained from murine bone marrow. Diabetes. 2004 Jul;53(7):1721-32.

[108] Achenbach P, Barker J, Bonifacio E. Modulating the natural history of type 1 diabetes in children at high genetic risk by mucosal insulin immunization. Curr Diab Rep. 2008 Apr;8(2):87-93.

[109] Monti P, Scirpoli M, Maffi P, Piemonti L, Secchi A, Bonifacio E, et al. Rapamycin monotherapy in patients with type 1 diabetes modifies CD4+CD25+FOXP3+ regulatory T-cells. Diabetes. 2008 Sep;57(9):2341-7.

[110] Simon G, Parker M, Ramiya V, Wasserfall C, Huang Y, Bresson D, et al. Murine antithymocyte globulin therapy alters disease progression in NOD mice by a time-dependent induction of immunoregulation. Diabetes. 2008 Feb;57(2):405-14.

[111] Xiu Y, Wong CP, Bouaziz JD, Hamaguchi Y, Wang Y, Pop SM, et al. B lymphocyte depletion by CD20 monoclonal antibody prevents diabetes in nonobese diabetic mice despite isotype-specific differences in Fc gamma R effector functions. J Immunol. 2008 Mar 1;180(5):2863-75.

[112] Ciancio G, Burke GW, Gaynor JJ, Roth D, Sageshima J, Kupin W, et al. Randomized trial of mycophenolate mofetil versus enteric-coated mycophenolate sodium in primary renal transplant recipients given tacrolimus and daclizumab/thymoglobulin: one year follow-up. Transplantation. 2008 Jul 15;86(1):67-74.

[113] Skyler JS, Krischer JP, Wolfsdorf J, Cowie C, Palmer JP, Greenbaum C, et al. Effects of oral insulin in relatives of patients with type 1 diabetes: The Diabetes Prevention Trial--Type 1. Diabetes Care. 2005 May;28(5):1068-76.

[114] Thivolet C. New therapeutic approaches to type 1 diabetes: from prevention to cellular or gene therapies. Clin Endocrinol (Oxf). 2001 Nov;55(5):565-74.

In: Genetic Diagnoses
Editor: R. J. Sarma, pp. 143-188

ISBN: 978-1-61324-866-9
© 2012 Nova Science Publishers, Inc.

Chapter VII

State of the Heart on Genetics of Congenital Heart Diseases: Molecular Basis, Genetic Diagnosis and Counselling

Giuseppe Limongelli[*], *Paolo Calabro, Valeria Maddaloni,*
Raffaella D'Alessandro, Giuseppe Pacileo and Raffaele Calabro
Monaldi Hospital, Second University of Naples, Naples, Italy

Abstract

Congenital heart disease is the most frequent form of major birth defects in newborns affecting close to 1% of newborn babies (8 per 1,000).

The etiology is multifactorial, including a genetic basis (causative genes, or interactions between genes and environment) and the influence of non-inherited risk factors (such as multivitamins, maternal illness, drug administration, environmental agents exposure or also maternal and paternal sociodemographic factors).

Heart is considered the first functional organ of the embryo. It develops from the mesodermal sheets through the formation of an early linear heart tube which begins to contract at the eight- to nine-somite stage before the formation of the chambers and the conduction system. Recently, several CHDs have been found to be caused by mutations of the genes involved in the heart embryogenesis (*TFAP2B, Tbx1, NKX2-5, NKX2-6, ZFPM2/FOG2, GATA4*). These genes are often investigated as candidate genes if their biological role could be associated to the cardiac defect.

Genome wide linkage analysis, candidate gene association studies, RNA expression profiling and resequencing are commonly used techniques to identify genes responsible of a certain cardiovascular disorder. The method of analysis can be selected on the basis of the known information about the disease.

[*] Address for Corrispondence: Giuseppe Limongelli, MD, Phd, EDBT, FESC, MAHA, Department of Cardiothoracic Sciences, Second University of Naples, Monaldi Hospital, Via L Bianchi, 80131, Naples, Italy, Email: limongelligiuseppe@libero.it. Work-phone:+390817062852, Mobile: +393381041147, FAX: +390817062683.

As a consequence of the increasing number of congenital/genetic cardiovascular diseases discovered in recent years, the genetic counselling has become very useful to inform the family of the affected subject about the hereditary risk, and to suggest genetic testing for family members.

In addition, due to the significant improvement of non invasive imaging techniques (ultrasound imaging) and molecular analysis, prenatal diagnosis is becoming available for many congenital disorders.

Chromosomal karyotype with increased band number to identify large chromosome rearrangements, gene or region specific FISH analysis for detecting deletions/insertions or aneuploidies, indirect PCR assays to assess small gene specific mutations or deletions/insertions, or direct sequence analysis for the detection of specific point mutations represent some of the available techniques to detect the disease in an early phase. Examples of genetic proved congenital disorders for which a genetic test is available are DiGeorge syndrome (deletion 22q11, *Tbx1* gene), Williams-Beuren syndrome (microdeletion 7q11.23, gene contiguous syndrome), Alagille syndrome (deletion 20p12, *JAG1* gene), Noonan and LEOPARD syndrome (*PTPN11, SOS1, KRAS,* and *RAF-1*), Holt-Oram syndrome (mutations in *TBX5* gene).

Keywords: congenital heart diseases, heart development, genes, molecular techniques, genetic counselling.

Introduction

Cardiovascular diseases (CVDs) and congenital heart diseases (CHDs) are the major health problems around the world [1, 2]. The frequency of cardiac malformation at birth is 4 to 10 liveborn infants per 1000 and 40% of them are diagnosed until the first year of life [2]. Some of the CVDs show a mendelian inheritance; examples are familial forms of hypercholesterolemia, hyperhomocystinuria, Hutchinson-Gilford progeria syndrome, Tangier disease, an inherited form of coronary heart disease and "channelopahties" (APPENDIX).

However, the most common forms of CVD are believed to be multifactorial, in fact it is supposed that many genes, each with a little influence on the final phenotype, could interact each other and with modifier genes or environmental factors. Gene variations mentioned before could also have a contribute in the predisposition to develop cardiovascular diseases, in fact the same genotype could be associated with different phenotypes because of the environmental influences (such as cigarette smoking, age and pharmacological treatment) [3].

Moreover, although there has been a long-standing clinical view that most CHDs occur as isolated cases, today we know that there is a genetic basis for several conditions. In fact, in recent years, through studies of recurrence and transmission risks, an hypotheses of genetic predisposition of the subject that interacts with the environment to cause CHD (multifactorial aetiology) has been proposed. These results have been confirmed by molecular genetic studies which have provided clear evidences for the genetic implication in some CHDs such as septal defects and patent ductus arteriosus. This development in the knowledge about the causes of CHD shows how genetic involvement in that class of pathologies has been underestimated in the past, although human cardiovascular genetics is only at the start point of investigation and the discoveries are in continuous updating [2].

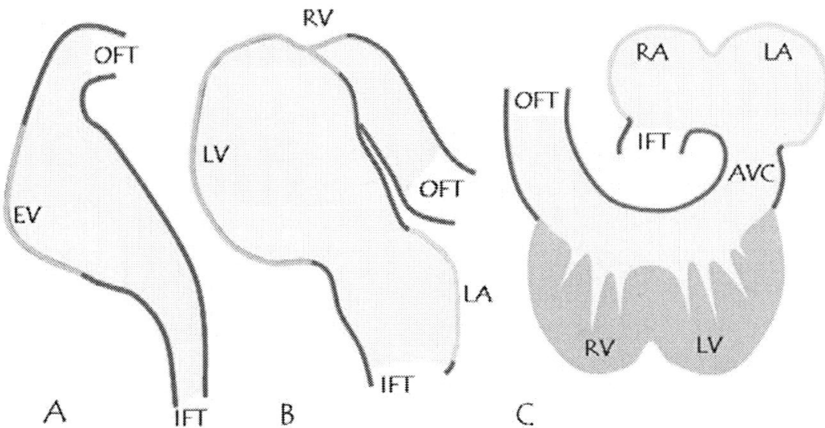

Figure 1. A: linear heart. B: looped heart. C: four chamber heart.

Not only genetic causes, but also non-inherited risk factors involved in the development of CHD have been widely investigated in order to prevent parental exposures to dangerous substances during the three months before pregnancy and the first trimester (periconceptional period) [4].

Although it has been a great development in the knowledge of the genetic basis of several cardiovascular diseases, for many other diseases are still unknown the genes involved in the disease development, today. It could be due to genetic heterogeneity, differing patterns of linkage disequilibrium in varying population groups, and phenotypic heterogeneity.

The multifactorial nature of many phenotypes is evident also in the individual response of some subjects to drugs, so the finding of genetic variations involved in disease development could be useful also to detect which genetic variant is causative of the failure to a certain drug therapy with a relevance for public and clinical health.

The pharmacogenetic approach is focused on the detection of variations of the receptors of the drug, the enzymes involved in metabolism and clearance of the pharmacological principle; the knowledge about the genetic basis of drug response is extremely interesting in the development of clinical therapy design in according with the needs of the specific patient [1].

This chapter deals with pathogenesis mechanisms and molecular basis underlying CHDs.

1. Embryogenesis of the Heart

Heart is considered the first functional organ of the embryo. It develops from the mesodermal sheets through the formation of an early linear heart tube which begins to contract at the eight- to nine-somite stage [5], before the formation of the chambers and the conduction system. The linear tube begins forming by the second week of human gestation and it is formed by a crescent of mesoderm tissue. After, the looping of the linear heart leads to the formation of separate compartments which are outflow tract (OFT) at one end of the looped heart [6], embryonic right ventricle (RV), embryonic left ventricle (LV), atria and sinus venous [5], while the inflow tract (IFT) is at the other end of the looped heart [6]; contemporary gap-junctions begin to appear. Subsequently occours the connection between the right atrium and the right ventricle and between the left ventricle and the outflow tract, respectively [5] (see Figure 1).

The primary heart, formed by a linear tube, loops and balloons to develop atria and ventricles (the OFT forms the aorta and pulmonary arteries while the IFT becomes the atrioventricular canal) [6]; these compartments are initially arranged in series and communicate via the atrioventricular canal and the interventricular foramen, this means that the right atrium and ventricle have not a direct communication point in the early phases of the heart formation [5]. The ballooning initiates when the neural crest cells migrate into the heart and there is an increase in the size of the individual chambers [6]. The final heart is then formed after the physical separation between right and left sides of the initially common atrium and the formation of the right atrioventricular connection. At first, the atrioventricular canal is present only above the left ventricle, while the right-sided parts of the junction is formed after the local expansion of the myocardium: the just formed structure shows typical features shared by atrioventricular and interventricular junctional myocardium. The development of the right ventricle is related to the rightward expansion of the atrioventricular canal. The final fibro-fatty atrioventricular junction is formed when the initial myocardium of the canal is sequestered as an atrial structure. This mechanism explains some cardiac malformations with the interruption of the development of some heart structures [7].

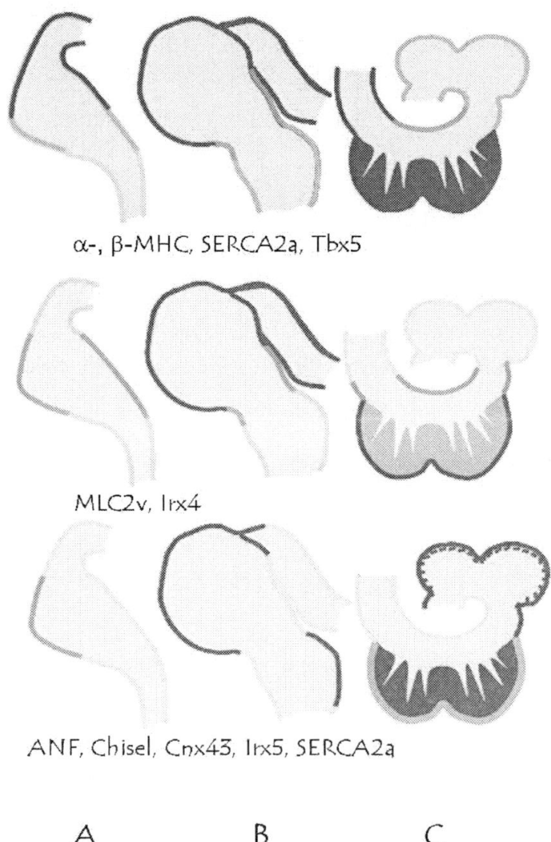

Note 1. For text abbreviations see Appendix II.

Figure 2. Gene expression profiles during heart embryogenesis. A: E 8.5-9. B: E 9-10. C: E 10-12.

Several gene expressions have been investigated to understand which factors are involved in heart development. There is a default pattern of expression in the linear heart that remains unchanged in IFT, AVC (atrioventricular channel), OFT and inner curvature during chamber formation; this pattern includes α-*MHC*, *β-MHC*, *Tbx5*. After the formation of the chambers, there is the activation of other genes such as *ANF*, *Chisel* and *Irx5* and the upregulation of *SERCA2a*, *MLC2v* and *Irx4* and it is the specific expression program of the tubular heart and the outer curvature of the looped heart. Differences have been found also in expression pattern along the anteroposterior (A/P) axis, with the steadily increasing of *HRT-1* and -2, Tbx5, *dHAND* and *GATA-4*; this polarity is likely at the basis of the correct heart structures localization, in fact the specific positions of heart parts must be defined by specific factors expressions. Moreover, *Irx4* and *Hand1* have been supposed to be the defining factors for the limits between the A/P and dorsoventral (D/V) regions, respectively. The unequal expression of these genes leads to the differentiation of heart regions, for example atria and ventricles have higher conduction velocities and density of gap junction, while the myocardium flanking segments retain a poor electrical coupling and a long contraction duration, specific characteristics of the linear heart tube (experiments on mammalian embryos) [5].

2. Techniques Used for the Discovery of the Genetic Basis of Cardiovascular and Congenital Heart Diseases

Molecular cardiology is the field of the cardiology interested in characterizing the genetic modifiers involved in complex congenital cardiovascular disease and specific cardiovascular genes causing monogenic disorders. The principal strategies used in gene localization and gene product analysis are two: functional and positional cloning. The first requests some information about the biochemical defect underlying the disease to identify the relative gene; it is not always applicable because often no information about the aetiology of the disease, especially when one deals with complex diseases due to the interaction of several genes. The second approach has become the most common method of disease-gene identification because it doesn't request any information about the processes at the basis of a certain disease. Moreover, the sequencing of the human genome project has led to a rapid increase in the data available about the gene composition of the human genome, making easier this approach; it is usually followed by biological assay for candidate gene role assessment.

Positional cloning recognizes a disease-gene through its position, it is necessary to have some information about the region where the gene is located (for example the finding of an affected patient with a chromosomal deletion); then, a physical or genetic map of the genomic region is derived. Both kind of maps are obtained through the identification of genetic markers, used as reference points on the genome; these have a different nature, indeed in earlier studies RFLP (restriction fragment length polymorphisms) were used, but then, the discovery of mini- and micro-satellites has allowed the construction of higher resolution maps.

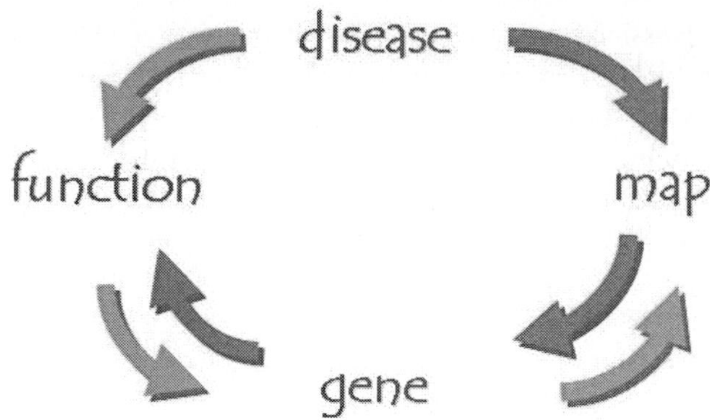

Figure 3. Functional cloning strategy is represented by green arrows and positional cloning strategy by blue arrows.

So, genetic markers are useful for positional assignment of a gene: markers will show co-inheritance (linkage) with the gene responsible for the phenotype if they are located in the same chromosomal region. Repeated inheritance of a marker in affected members of a family will indicate, with high probability, that the gene is near to them; conversely, if a gene is distant from some markers, their segregation will appear independent from the disease. Finally, a gene is identified in the found region if some genes are already known to belong to the genomic fragment, or by sequencing the region and comparing its sequence to known genes sequence [8].

An example of identification of genes related to an inherited (and, sometimes, congenital) cardiovascular disease is given by hypertrophic cardiomyopathy (HCM). Its cause has been found to be mutations in α- and β-myosin heavy chain (MHC); indeed, positional cloning has allowed the identification of a genomic region linked to HCM. *MHC* genes were already known to be located within this region and their biological functions were consistent with the development of the disease as they are expressed at high levels in the myocardium and encode a major contractile component of myofibrils. Mutations in such genes can disrupt the protein structure and cause the lack of the correct organization of cardiac fibrils [9].

A further explanation of the techniques used in genetic studies is reported.

Linkage Analysis

Genome wide linkage studies are performed using genetic and phenotypic data from families with recurring diseases. In this approach it isn't necessary to have genetic information about the disease because the method allows to identify genomic regions eventually involved in the development of the disease. In fact, several genetic polymorphic markers (located in known regions of the genome) are usually tested to find if some of them are inherited in association with the pathologic features; if any marker shows co-segregation with the disease, the genomic region in which it is located is analyzed, especially if it is known to contain any candidate gene for the disease. The reason why markers could be associated to the pathologic features resides in the very low frequency associated with a crossing over event between two close genetic loci, so, once the marker of the region involved in the disorder is identified, it is likely that the disease-causing gene is in the same region and that they are inherited together (they are said to be "linked"). The genetic, phenotypic and pedigree data are then analyzed by a specific software which calculates the degree to which the marker information is identical by descent among family members in the pedigree and how this degree of genetic similarity for a marker correlates with the phenotypic resemblance among family members.

A LOD (logarithm of the odds) score is associated to each marker investigated and it represents the probability that it and the region in which is located the disease-causing gene are linked. If the statistical analysis gives a positive result (a LOD score >3, between 2 and 3, is estimated of significant linkage, because the likelihood of the finding of that result with the non-association of the marker and the causative locus is <1 in 1000), a positional approach to identify which gene of the selected genomic region is responsible of the phenotype is usually performed, giving priority to the genes known to have a biological role related to the features of the disease [10].

Once identified the genomic region of interest, it is possible to intensify the resolution by adding markers (known to be located in that region) and repeating linkage analysis to have a set of fine mapped genotypes. It allows to confirm or not the early hypothesis of association between the region and the disease and to identify more clearly which gene is involved in the disease [1].

Candidate Gene Association

It is possible to perform a candidate gene association only when the genetic region involved in disease development is known (for example after linkage analysis) or when there is some candidate gene because of its biological function. So, this approach allows the study of a single polymorphism or a set of them on a chromosome (haplotype) to investigate the possibility of a common inheritance. The frequencies of the studied alleles or haplotypes are tested in a case and a control group and then they are compared to observe eventual variations between them and evaluate the possible implication in the disease. The identification of an association between a gene or an haplotype and a trait has not to be interpreted as an assumption of a causative role of the gene; in fact the allele or the haplotype could also be only in linkage disequilibrium with another unidentified genetic variation which is causative of the phenotype [11].

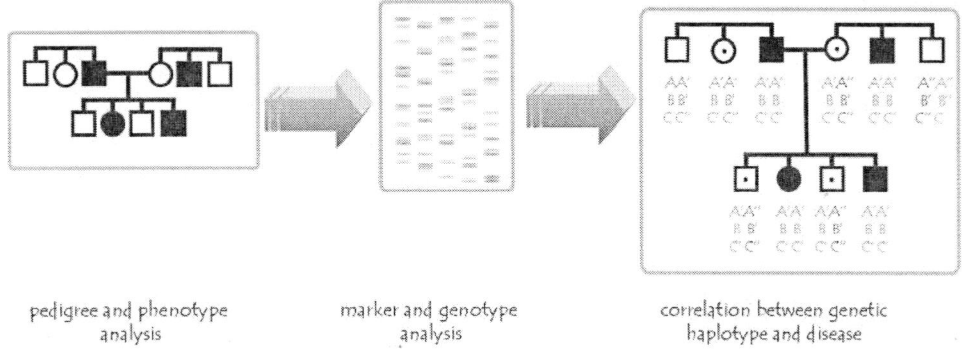

Figure 4. An example of linkage analysis.

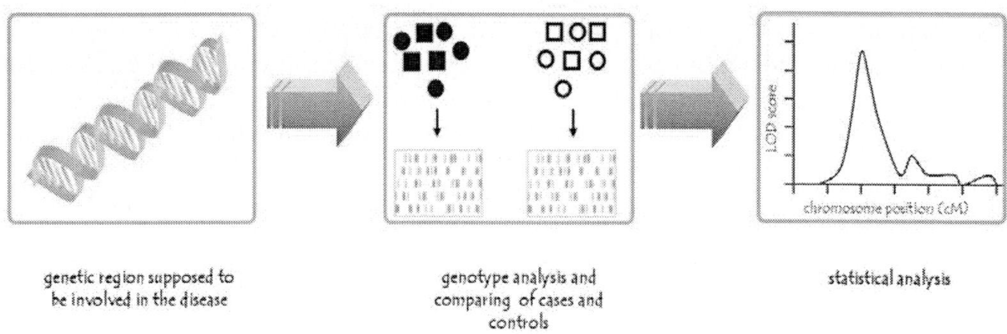

Figure 5. An example of gene association study.

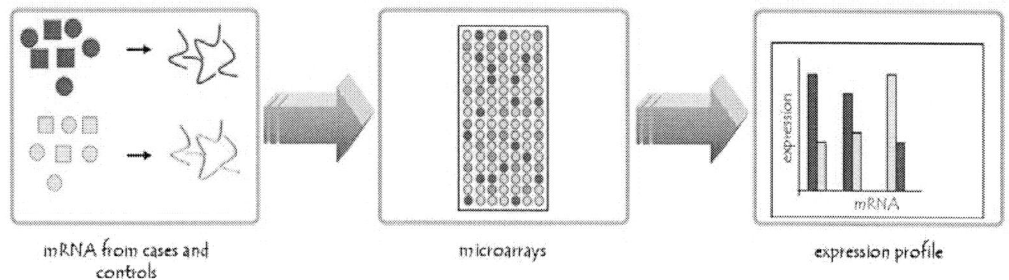

Figure 6. Scheme of a mRNA expression profile study.

RNA Expression Profiling

Another possible initial approach is the study of the differences between the expression profiling of thousands of genes affected and normal subjects. It can provide information about which gene could cause the pathologic condition and so help the selection of the gene to study [1].

In fact, not only mutations in the coding region of a gene, but also mutations in the next regulating regions can determine a different phenotype; in the first case, modifications of the sequence specifying for aminoacids can modify the protein primary structure (through a missense mutation), determine the interruption of the protein synthesis (if a codon for an aminoacid is replaced by a stop codon), or modify the splicing pattern (if the genetic variation creates or removes a sequence involved in the splicing process); all these modifications determine the production of a protein with different three-dimensional structure, different chemical and physical features an biological properties. Otherwise, mutations in the regulating regions at the 5' or the 3' end of the gene, can affect the initiation and the rate of the transcription or of the translation and the stability of the mRNA; these other mutations determine a quantitative difference in protein production. So, the detection of different RNA expression profiling can be useful especially in determining the last class of modifications, in fact a disease could also be due to an imbalance in the production of a specific protein.

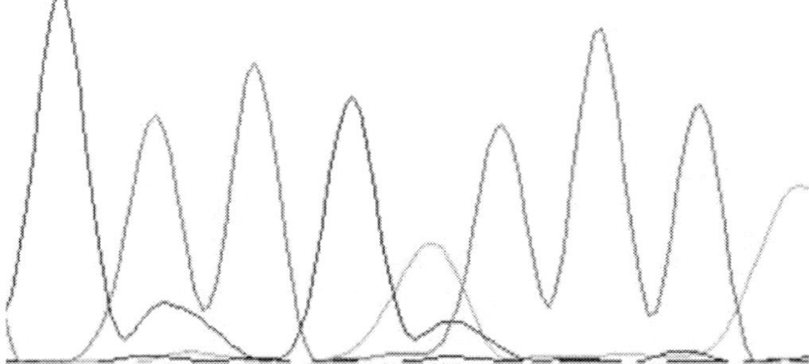

Figure 7. A resequencing electropherogram.

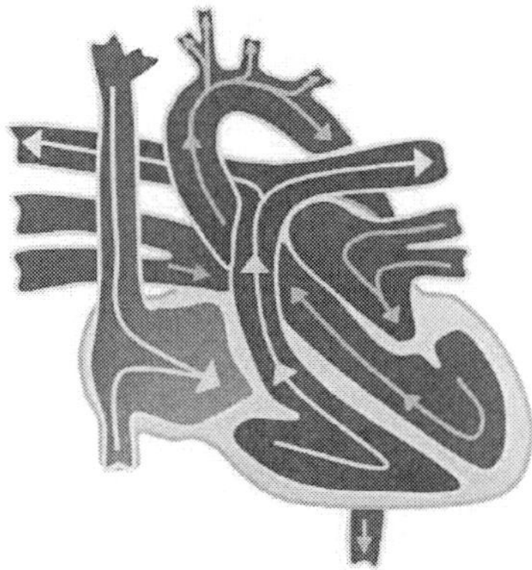

Figure 8. An healthy heart.

Resequencing

Resequencing is a useful technique performed with the aim of detect less frequent genetic variations. In fact common indirect genetic tests search for very frequent mutations, and do not detect the eventual presence of other new or never detected before genetic variations. Also if whole-genome resequencing is excessively expensive, several studies designed for specific candidate gene, consent the detection of new or rare variants by resequencing [1].

3. Genes Involved in Heart Development

The study of genetic heart diseases have led to the comprehension of the factors involved in heart development and their mechanism of action through the observation of the type of inheritance.

TFAP2B

It is a transcription factor expressed in neuroectoderm (neural crest cells are important in the development of several of the tissues affected in Char syndrome(characterized by patent ductus arteriosus, PDA, facial and hand anomalies); among them is included also the cardiovascular tissue, particularly conotruncal septation) [12]. *TFAP2* genes belong to a family of closely related and evolutionarily conserved sequence-specific DNA-binding proteins, identified in *Drosophila melanogaster* (invertebrates), *Xenopus laevis*, chickens, mice and humans (vertebrates). *Drosophila* has only one *TFAP2* gene, but evidence for evolutionary gene duplication is given by the presence of at least three genes in mice and human (*TFAP2a, TFAP2b* and *TFAP2c* in mice and *TFAP2A, TFAP2B* and *TFAP2C* in human) [13]. *TFAP2B* gene is supposed to be responsible of several defects characterizing Char syndrome [6]. The resulting proteins form homo- and heterodimers which bind GC-rich sequences in the promoter regions of certain genes, activating their transcription. *TFAP2B* has been found associated to Char syndrome in humans, but the correspondent genes in *Drosophila* and mice are also responsible of phenotypic anomalies. In Char syndrome has been found that one missense mutation (P62R), altering the HSH domain necessary for DNA-binding, is particularly recurrent in patients with PDA (patent ductus arteriosus) (in foetal life the ductus arteriosus, a muscular artery, shunts blood from the pulmonary artery to the aorta, bypassing the lungs). Its abrupt closure at birth establishes the mature circulatory pattern and represents a dramatic example of vascular remodeling. Failure of this normal process results in persistent PDA, which left untreated can result in pulmonary hypertension and heart failure [13], mild facial anomalies and normal hands. Although the protein is able to bind DNA, this mutation causes the failure in transactivation as a homodimer and interferes with transactivation from heterodimeric partners. It has been supposed that the discrepancy between the occurrence of PDA (very high in carriers) and craniofacial and hand anomalies (mild phenotypes) was due to a differential expression pattern of TFAP2 coactivators [14]. So, PDA in Char syndrome is likely to result from abnormal neural crest development. The explanation of this may be the derivation of the left ductus arteriosus from the neural crest cells, while associated structures as pulmonary artery and descending aorta do not show the same derivation. As *TFAP2B* has been demonstrated to be expressed in neuroectoderm, its mutations could affect the constriction or degeneration of the ductus after birth, but not perturb the adjacent arterial structures [12].

Figure 9. Patent ductus arteriosus.

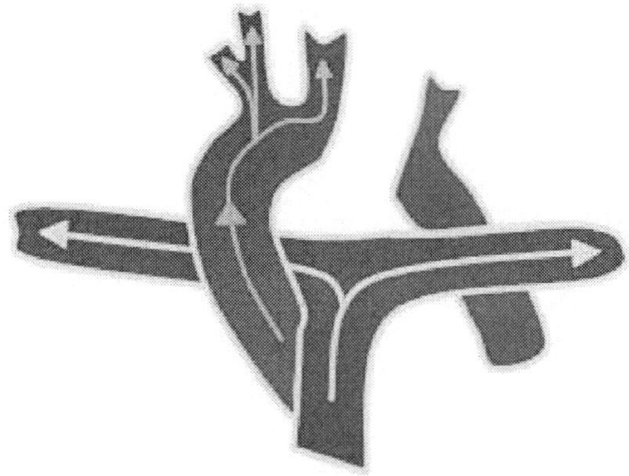

Figure 10. Interrupted aortic arch, type B.

Another genetic region showing association with the presence of PDA (in non-syndromic cases) is 12q14. It spans over 2.8 million bp and contains 28 known genes and 7 hypothetical genes. So, there isn't an identified gene yet, but a set of possible causing genes [13].

Tbx1

Tbx1 is a T-box putative transcription factor, required for the segmentation of the pharyngeal apparatus, and homozygous mutations lead to cardiovascular, craniofacial, ear, thymic and parathyroid defects. It has been supposed to be one of the genes involved in the development of DiGeorge syndrome because its haploinsufficiency causes aortic arch defects in mice similar to the deletion of the genomic region 22q11, cause of the DiGeorge phenotype; moreover the 22q11 region includes the *Tbx1* gene, so it's a candidate for causing that disease [15]. It has been found that some patients with DiGeorge syndrome lacked the 22q11 deletion but had mutations in *Tbx1* gene [6]. The haploinsufficiency in mouse leads to remodelling defects of the fourth pharyngeal arch arteries (PAAs), and then to interruption of the aortic arch type B (IAA-B). However, phenotypic features of patients affected by DiGeorge syndrome are very variable, while *Tbx1* mutation appear causing homogeneous defects; an explanation could be a variable expression or a reduced penetrance in carriers, due also to interactions between *Tbx1* and other genes. It has been showed that mutations of *Fgf8* (Fibroblast growth factor 8) can affect the phenotype associated with *Tbx1* mutations, enhancing the effect on the fourth PAA, although the eventually interaction has not been characterized yet. It is possible that *Tbx1* interacts with *Fgf8* promoter, but it is not sure because the genetic elements bound by the transcription factor are not well known, now; moreover, the effect of the interaction would be restricted to the pharyngeal endoderm since the expression pattern overlaps only in that tissue and it is confirmed by the interaction between the two genes in the development of the aortic arch and thymus [15].

Tbx1 has also been demonstrated to interact with the transcription factor Sonic hedgehog (Shh) and Fox proteins. *Tbx1* gene, in fact, contains a Fox binding protein, essential for gene expression in the pharyngeal endoderm regulated by Foxa2. This protein interacts with GATA-4 transcription factor to relax chromatin, and so it could have a general role in activation of the transcription; moreover it is possible to suppose a similar interaction between Fox a protein and GATA-3 transcription factor in the regulation of *Tbx1* and it could explain the correlation with DiGeorge syndrome (GATA-3 is linked to a second DiGeorge syndrome locus on 10p). This pathway involves an indirect control of Shh on *Tbx1* expression through the intermediate action of Fox proteins, particularly Foxa2 protein in the pharyngeal endoderm (Foxa2 expression is dependent on Shh signalling). Due to this interaction, patients affected by DiGeorge syndrome, but without any mutation in *Tbx1* coding sequence could be genetically investigated for eventual mutations in the regulatory *cis*-element Fox-binding or in the *trans*-activating elements (Fox proteins or Shh) [16].

NKX2-5, 2-6

NKX2-5 is a homeobox transcription factor involved in heart development. Indeed, subjects with *NKX2-5* mutations show congenital heart defects including ventricular septal defects and Tetralogy of Fallot; the variations may occur most of the times in the region coding for the homeodomain and sometimes in the 5' and 3' end of the gene and are responsible for the presence of a truncated protein with a dominant negative effect or deleterious gain of function. This finding suggests that the transcription factor is involved in several mechanisms of the heart formation such as atrial, conotruncal and ventricular septation, AV conduction and AV valve formation. Moreover, in subjects affected by conotruncal defects without a deletion 22q11, a mutation in *NKX2-5* gene is present [17] and in atrioventricular conduction block with or without associated CHDs. Moreover, carriers of *NKX2-5* mutations are thought to have a lifelong risk of development of heart block and sudden death [18]. Both familial cases and sporadic cases (in which the proband had a *de novo* mutation) have been described. The way with the *NKX2-5* mutation can affect heart development is not clear yet, but it is known that the transcription factor activates several other genes involved in heart formation as *Hand1* (heart and neural crest derivatives expressed 1, a transcription factor), myocyte enhancer factor-2 (*MEF2*), myosin light chain 2V (*MLC2V*), brain natriuretic peptide (BNP), atrial natriuretic factor gene, cardiac ankyrin repeat protein gene, *N-myc*, and *MSX2* (msh homeobox 1, a transcription repressor). Nkx2.5 binding to target DNA may be dependent on interaction with GATA-4 and it suggests another way in which mutations may alter the specificity and affinity of DNA binding, disrupting regulated gene expression that is essential during cardiac development [17].

The mutations related to Tetralogy of Fallot (found in about 4% of affected nonsyndromic subjects) do not disrupt the homeodomain sequence (however occur in conserved positions changing the aminoacidic residue) and are not fully penetrant, as demonstrated by its presence also in non affected relatives of the probands; this suggests an eventual implication of environmental other than genetic factors [19].

NKX2-6 is a transcription factor similar to *NKX2-5*. They share, in homeodomain regions, 90% of identity and their expression pattern is overlapping. A *NKX2-6* missense mutation (F151L) has been found associated to a familiar case of persistent truncus arteriosus; it occurs in the homeodomain sequence and impairs DNA binding and transcriptional activity. It is expressed in caudal pharyngeal arches and at the opposite poles of the developing heart. While *NKX2-5* loss in mouse has shown embryonic lethality, *NKX2-6* mutations are neither lethal nor lead to cardiac malformation, because *NKX2-5* expression extends also to the pharyngeal endoderm with a compensation mechanism. It is not equally for all *NKX2-6* mutations, but only for a specific variation (not for T451C, but for F151L) probably because the second one form a less stable protein and *NKX2-5* is not completely able to provide for *NKX2-6* functions. Alternatively, the F151L mutant could have a dominant negative effect not yet identified. Finally, the mechanism of action of the mutant protein could be understood after the identification of interactions between NKX2-6 and other transcription factors as GATA-4 and TBX, similar to the action NKX2-5 [20].

ZFPM2/FOG2

ZFPM2/FOG2 is a protein expressed during early heart development and contains eight zinc finger motifs. It associates with GATA-4 transcription factor and acts as a coregulator. The identification of congenital heart defects in mice with mutations of that gene has led to the investigation of it also in human patients affected by Tetralogy of Fallot. There have been found two missense mutations in nonsyndromic patients affected by TOF: E30G and S657G. Neither the first mutation nor the second one are in the regions coding for the zinc finger domains, but S657G shows a reduction in repression of *GATA-4* mediate transactivation of the BNP promoter. Maybe the influence of the mutation is given by the incorrect folding of the protein or by a possible phosphorylation of the Serine residue that is impossible when it is replaced by a Glycine [21].

Figure 11. Ventricular septal defect.

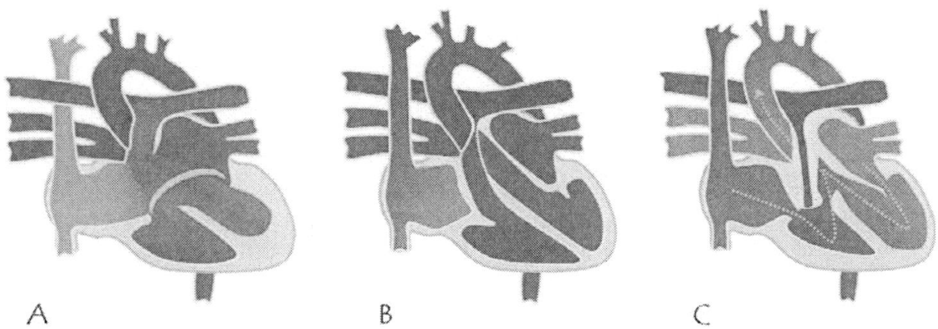

Figure 12. Congenital heart malformations. A: atrioventricular septal defect. B: pulmonary valve stenosis. C: tetralogy of Fallot.

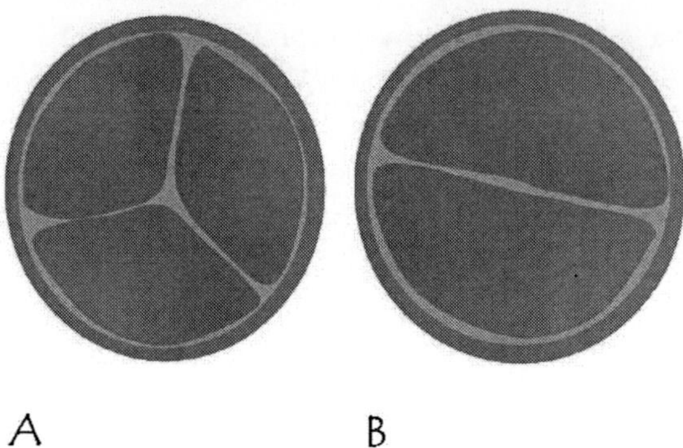

Figure 13. A: tricuspid aorta valve. B: bicuspid aorta valve.

GATA-4

GATA-4 (localized at chromosome locus 8p23.1) gene encodes a transcription factor known to play an important role in cardiogenesis. Mutations have been found in patients with isolated autosomal dominant ostium secundum atrial septal defects (ASD) and in families with isolated CHDs. However, other mutations have been identified in subjects with ventricular septal defects (VSDs), atrioventricular septal defects (AVSDs), pulmonary valve stenosis and recently in patients with tetralogy of Fallot (TOF). It has been suggested that *GATA-4* implication in CHDs could be only modest, since not all subjects with ASD show *GATA-4* mutations [18].

4. Non-Inherited Factors Involved in CHD

A recent discovery is the possibility of the multivitamins containing folic acid to prevent, as well as neural tube defects, CHD as conotruncal defects, ventricular septal defects (VSDs), while women that used medications containing folic acid antagonists had an increased risk for their child to develop CHD. Also maternal illnesses and conditions can influence the risk of developing CHD in their child. Some well known pathologies are phenylketonuria (associated with a sixth-fold increased frequency of heart defects), overall Tetralogy of Fallot, VSDs, patent ductus arteriosus (PDA) and single ventricle; but the risk rate can be reduced by diet control before and during pregnancy; maternal diabetes is associated with laterality and looping defects, transposition of the great vessels, VSDs, hypoplastic left heart syndrome, conotruncal defects, outflow tract defects, cardiomyopathy and PDA; HIV can be transmissed vertically from the mother to the offspring and children infected in uterus have an increased risk of dilated cardiomyopathy and left ventricular hypertrophy; offspring of women with epilepsy have an increased risk for congenital malformations. Several drugs can also have an effect on the foetus; indeed it has been investigated the effect of maternal therapeutic drug exposure of thalidomide which is a cardiac teratogen involved in the development of various cardiovascular defects; vitamin A cogeners administering during pregnancy is cause of central nervous system malformations, micrognatia, cleft palate, thymic and eye anomalies and cardiac and great vessel defects; a slightly higher rate of exposure to phenylephrine, corticosteroids, folate antagonists, female hormones, narcotics and ACE inhibitors during pregnancy has been observed in mothers of children with congenital heart diseases; the maternal use of ibuprofen during pregnancy has been found as an exposing risk to the development of dextro-looped transposition of the great arteries, membranous VSDs, ASDs, Down syndrome, bicuspid aortic valve. Also some non-therapeutical drugs could have an effect on the development of CHDs. Among them there are caffeine which is related to the increasing of the risk of the offspring to have congenital anomalies; alcohol has documented teratogenic effects, including cardiac malformations; cocaine ingestion during pregnancy induces coronary occlusion in the developing foetal heart. Moreover, the exposure of the mother to environmental agents such as herbicides (it has been found associated to the increased risk of conotruncal defects), air quality (the increased level of carbon monoxide stimulates the presence of VSDs, while high levels of ozone lead to aortic artery and valve anomalies). Finally, it has been investigated if socio-demographic characteristics of the parents could have an effect on CHDs in children: an example is the ethnicity, in fact, white infants have been found to have an increased prevalence of Ebstein's anomaly, aortic stenosis, atrioventricular septal defects, ASDs, coarctation of the aorta, truncus arteriosus, transposition of the great arteries, Tetralogy of Fallot, PDA, hypoplastic left heart syndrome, and a decreased prevalence of pulmonary stenosis; a history of reproductive problems has been associated with an increased risk of Tetralogy of Fallot, non-chromosomal atrioventricular septal defects, ASDs and Ebstein's anomaly; maternal stress, measured by maternal reports of job loss, divorce, separation, or death of a close relative or friend, was found to be associated with an increased risk of conotruncal heart defects; paternal age was found associated with CHDs in the offspring, indeed risk for men >25 years of age also was increased for VSDs, PDA, and tetralogy of Fallot [4].

5. Pre- and Post-Natal Diagnosis

The increasing amount of information about the genetic basis of CHDs has led to the development of genetic testing of embryos, foetuses, children and adults both for research and clinical purposes.

The discovery of the genetic basis of CHDs allows the clinicians to have more prognostic information for clinical outcomes, to know genetic reproductive risk and inform the other members of the family of it, to test other family members if it is required.

Nowadays, there are some genetic tests available for the diagnosis of genetic alterations in patients with CHD (both children and adults); they consist in cytogenetic or DNA mutation analysis.

The first kind of analysis is very common and allows the detection of numeric and structural chromosomal variations (the second one only to identify small arrangement variations).

Chromosome analysis have revealed aberration in 8% to 13% of neonates with CHD and at least 30% of children with chromosomal abnormalities have a CHD. The techniques used to detect chromosomal defects are standard metaphase or prometaphase karyotype; the difference between them is the resolution of the stained bands, indeed the first one can show 450 to 550 bands, while the second one can show 550 to 850 bands. The common use of cytogenetic analysis is due to the possibility of performing both pre- and post-natal diagnosis thanks to the facility of sample collection (peripheral blood lymphocytes, cord blood, amniotic fluid, chorionic villi). Moreover, the introduction of the FISH (*fluorescent in situ hybridization*) technology has led to the increase of sensitivity with the use of gene-specific probes. These analysis are particularly appropriate for the detection of aneuploidies, deletions, inversions, translocations or duplications, but to identify little changes such as variations in the sequence of a single gene (both in the coding and in the regulation regions), there are some genetic tests that show small deletions, insertions or substitutions of the nucleotides. These tests can be based on polymerase chain reaction (PCR) assays to have an indirect diagnosis (examples are high-performance liquid chromatography and single-strand conformation polymorphism) or can perform a direct sequence analysis. DNA mutation analysis are very useful to find mutations in regulatory regions (like variations in the splicing pattern, in the promoter region) and also in the coding sequence (even mutations that don't change the aminoacidic sequence of the coded protein) but they have many limits because of the impossibility of detecting large gene and chromosome mutations [22].

6. Congenital Heart Diseases Detected by Genetic Tests

6.1. Digeorge Syndrome

DiGeorge syndrome (DGS) was described in 1965 as congenital absence of the thymus and parathyroid glands found at necropsy in three children [23]. In subsequent years the syndrome was described as characterized by aplasya of the tymus and of the parathyroid glands, cardiac malformations (development defect of the third and the fourth branchial arches) [24] and facial dysmorphia (hypertelorism with short palpebral fissures, small mouth and short philtrum, retrognatia and low set, posteriorly rotated ears) [23]; CHDs occur in about 75% of the patients with the syndrome [18] (type B interrupted aortic arch occurs in over 15% of cases of DiGeorge syndrome, truncus arteriosus in 7% of cases and Tetralogy of Fallot in 22%, see table for details). CHDs are the major cause of mortality (more than 90% of all deaths) in people affected by DGS [25].

DGS is a familial disorder inherited as an autosomal dominant trait and an average of 25% of the DGS patients show familial transmission of the following described deletion [26].

In early studies, a part of the patients with DiGeorge disease have been described to carry a deletion in the proximal long arm of 1 copy of chromosome 22 (22q11) [27].

The early studies performed to identify the cause of the disorder have led to the discovery of an unbalanced translocation with the loss of the short arm and proximal long arm of chromosome 22 (22pter→q11), resulting in monosomy for that region, while the remaining long-arm material (22q11→qter) is translocated to a number of different autosomes (exactly, there have been reported few patients with a 22q11.21→q11.23 deletion). At first, it has been supposed a contiguous gene syndrome, but recently some studies suggest that *Tbx1* is a possible candidate gene [15].

Table 1. Frequency of different congenital heart defects in DiGeorge syndrome affected patients. Other stands for hypoplastic left heart syndrome; pulmonary valve stenosis; double outlet right ventricle/interrupted aortic arch; bicuspid aortic valve; heterotaxy/A-V canal/interrupted aortic arch

CONGENITAL HEART DEFECT	AFFECTED PATIENTS
Tetralogy of Fallot	22%
Interrupted aortic arch	15%
Ventricular septal defect	13%
Truncus arteriosus	7%
Vascular ring	5%
Atrial septal defect	3%
Aortic arch anomaly	3%
VSD; ASD	4%
Other	4%

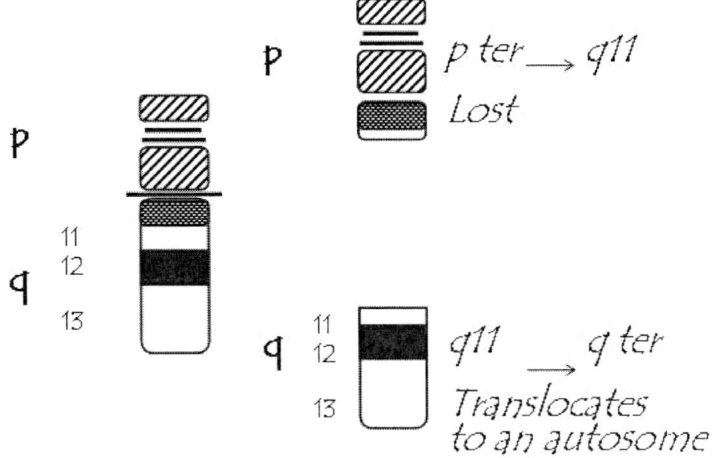

Figure 14. Genetics of DiGeorge syndrome.

However, recent high resolution techniques have allowed the identification of submicroscopic molecular deletions or small interstitial deletions; both are difficult to detect with classical cytogenetic methods because of the small size of chromosome 22 and the relatively uniform G-negative appearance of this band [28].

Some correlations between the genetic findings of similar diseases have led to the comprehension of possible mechanism that cause the development of the clinical phenotype.

In 1996, it has been identified the shortest overlapping deleted region between patients with DiGeorge and CATCH22 (CATCH is the acronym of cardiac defects, abnormal facies, T-cell immunodeficiency, cleft palate and hypocalcemia) syndromes; since the two diseases share some phenotypic features, the identification of the minimal deleted region in both syndrome affected subjects, is supposed to contain the genes responsible for the expression of the phenotype. The corresponding region is about 500 base pairs long. It has been found a sequence (possibly a gene), *DGCR5*, in which is contained the breakpoint; although there have been found more than one splice variant, it is possible that they belong to a class of

RNAs transcribed but not translated. The non-functional transcripts may have an indirect regulatory role, since *DGCR5* could be a *cis*-acting element which establishes the chromatin structure through transcription; in this way it can regulate proximal gene activation. This could explain the clinical features of the affected subjects: the disruption of *DGCR5* in patients can lead to deregulation of other genes [29].

Also velo-cardio-facial syndrome (VCFS) is characterized by a clinical set of symptoms similar to DiGeorge syndrome: cleft palate, cardiac anomalies, typical facies and learning disabilities; furthermore, the 85% of VCFS patients studied for genetic mutation assessment, show hemizigosity for 22q11.2 region; this allows to suppose that haploinsufficiency is the cause of VCFS, such as in the case of DiGeorge syndrome. Transcription mapping of the deleted region has led to the identification of several transcripts, including *N25-wa* encoding for CLTL, a protein homologous to clathrin heavy chain. It is one of the principal structural components of coated pits and coated vesicles, and has a main role in mediating intracellular vesicular transport, uptake membrane-bound ligands and fluids from the extracellular environment. In DGS and VCFS, the absence or reduction of that protein can affect the signalling necessary to the neural crest cell migration.

Another identified gene of the region is *DGCR2/IDD/LAN*, whose correspondent protein product is predicted to be an integral membrane protein. Its role is supposed to be interacting with other cells or extracellular matrix in the stabilization of cell interactions via ligand binding, and to have a possible role in neural crest cell migration; in fact, it has an N-terminus protein region containing Cys-rich repeats, similarly to the low-density lipoprotein receptor and basement membrane proteins.

Furthermore, other two transcripts have been found to be homologous to known genes: CTP shows 98% of identity to a rat mitochondrial tricarboxylate transporter (citrate transport protein); DGS-G shows 94% of identity with a mouse serine/threonine kinase (TSK-1).

CTP exchanges a tricarboxylate along with a proton for another tricarboxytlate/H$^+$ or a dicarboxylate or a phosphoenolpyruvate across the inner mitochondrial membrane and reduced levels of CTP due to haploinsufficiency may play a modifying role in DGS/VCFS by affecting glucose metabolism.

TSK-1 is a testis-specific transcript, but its function and its expression in embryos is still unknown, but the comparison with other Ser/Thr kinases has allowed the inclusion of TSK-1 in SNF1 subfamily of Ser/Thr kinases, involved in lipid metabolism, in establishing polarity in early *C.elegans* embryos and in early expression in the myocardial cells of the developing mouse heart. As protein kinases have a central role in coordinating the eukaryotic cell's response to external and internal signals, it is likely that DGS-G represents a candidate for DGS/VCFS development.

In non-deleted patients it is possible to suppose small or point mutations in these genes not yet identified [30].

As there is a high similarity among DGS, VCFS and CATCH 22 syndrome, it has been proposed that DGS is the severe end of the clinical spectrum of some other diseases [26].

Some possible mechanisms which cause the loss of genetic material could be unequal crossing over between multiple members of chromosome 22- specific gene families (leading to the production of de novo interstitial deletions), gonadal mosaicism for deletion-bearing chromosomes, cryptic balanced translocations between 22q11 and another acrocentric chromosome could arise during meiosis (in this cell cycle phase, all five pairs acrocentric chromosomes coalesce around the nucleous); all these mutation mechanism can lead to monosomy for 22pter→q11 and trisomy for pter→q11 of the other involved acrocentric chromosome [31].

Not all the patients with 22q11 deletion share the same clinical features, but the most common are cardiovascular anomalies, feeding, speech and learning disabilities and renal disorders [32].

In many cases, when the affected patient showed that deletion, also one of the parents was found to have it, but had no diagnosis of the syndrome because the clinical features were not very evident; the finding of this situation can help patients with non-evident clinical features to foresee the risk of the transmission of the mutant chromosome to the next generation. This is one of the reasons why sometimes is necessary to perform genetic analysis in familial genetic syndromes [18].

Genetic Test

The 15% of cases of 22q11 deletion is recognizable with a simple karyotype analysis, but with a more selected detection (such as FISH metaphase or interphase analysis) about the 95% of the cases can be diagnosed; some diagnostic laboratories perform deletion/duplication analysis of the interested gene [18, 25].

Prenatal testing is recommended for high-risk pregnancies; it is performed on foetal cells obtained by amniocentesis or corionic villi sample (respectively at 15-18 and 10-12 weeks' gestation) by FISH analysis (especially if ultrasound examination or echocardiography show the clinical features of the syndrome in the foetus). It is suggested also in low-risk pregnancies, when no family history is indicative of a possible inherited transmission of the mutation, but prenatal analysis reveal congenital heart diseases (such as conotruncal cardiac anomalies) or cleft palate. Nowadays, fewer than 5% of individuals with clinical symptoms of the 22q11.2 deletion syndrome have normal routine cytogenetic studies and negative FISH testing [25].

Genetic Counseling

Genetic counseling might be performed to assess the risk of transmission to the next generation; the best time to consider it is before pregnancy. It allows also the identification of eventual asymptomatic family members carrying the mutation.

The 22q11.2 deletion is inherited in an autosomal dominant fashion, but genetic transmission happens only in 7% of the cases as 93% are represented by *de novo* mutations in the foetuses. Anyway, for inheriting from an affected parent, the 50% of possible transmission is predicted [25].

Some indications for the choice of which patients should be tested for 22q11 deletion are given by the frequency of the genetic variation in some cardiac defects: it is suggested to perform a FISH analysis to all children with interrupted aortic arch type B or truncus arteriosus. Instead, the genetic testing of patients with tetralogy of Fallot is reasonable only after the detection of hypocalcemia, typical facial features, palate anatomy, speech and learning disabilities and endocrine abnormalities [33]. All these clinical features are evident in a child or adult patient, but to assess the deletion in the foetus, the only recognizable signs of the syndrome could be the cardiac anomalies [2].

The same genetic region is involved in another syndrome, the Cat-Eye Syndrome. In this disease, 22pter-22q11.2 is not deleted, but duplicated, indeed cytogenetic analysis show an inverted duplication of the short arm and proximal long arm of chromosome 22 (inv dup 22pter-22q11.2). The clinical features of the syndrome are very variable, but affected subjects often show ocular coloboma (of the iris and/or retina), anal atresia, preauricular skin tags and pits, heart defects (especially total anomalous pulmonary venous return), dysmorphic features (hypertelorism and down-slanting palpebral fissures, urogenital defects, mild to moderate mental retardation).

Since Cat-Eye syndrome and DiGeorge syndrome map to the same region, it has been supposed that gene(s) deleted in the second one is (are) responsible for at least some of the features of the first one. It has been supposed that the supernumerary chromosome could be an isodicentric chromosome, but marker loci analysis of the region has showed only trisomy and not tetrasomy (as attended) for some markers of the interested region; this demonstrates heterogeneity of the breakpoint in different subjects [34].

Williams-Beuren Syndrome

Williams-Beuren syndrome is an autosomal dominant disease that involves several body systems such as cardiovascular defects (supravalvular aortic stenosis, supravalvular pulmonary stenosis, peripheral pulmonary stenosis), hypercalcemia, defects of the skeleton and of the kidneys, elfin facies and cognitive and social disorders; moreover, half of the adult patients develop systemic hypertension [35].

About 80-90% of affected patients have cardiac anomalies [18]. Elastin arteriopathy has been found in about 75% of affected subjects and the most common form is supravalvular aortic stenosis (SVAS); peripheral pulmonary stenosis (PPS) is common in infancy; hypertension is common in adolescent and adults. Indeed individuals with SVAS and PPS are exposed to the risk of developing biventricular hypertrophy and hypertension with an increased risk of sudden death (its incidence about 1/1000) [25].

Genetics of Williams-Beuren Syndrome

The genetic cause of the syndrome was found to be a microdeletion (contiguous gene-deletion) at chromosome 7q11.23 in which is located also the elastin gene that causes the arterial abnormalities. The correspondent phenotype is variable depending on the deleted region. The genetic mutation that causes this disease is not inherited, but often it has been found arising *de novo* [35]. The deleted segment is about 1.5-1.7 Mb long and contains at least nineteen genes [36].

The disease is often caused by heterozigosity for a partial deletion of chromosome band 7q11.23. The known loci included in the deleted region encode for elastin (ELN), replication factor C subunit 2 (RCF2), LIM-kinase-1 (LIMK-1), wnt receptor *Drosophila frizzled* homolog FZD3, WBSCR1 and syntaxin1A (STX1A).

Only the 4-5% of patients which have been diagnosed as WBS cases show no deletions for any of the available markers; the majority are sporadic cases and rarely have been reported affected parents and children, indicating a high mutation rate (incidence of 1 in 20000 live births). The great number of spontaneous mutations is probably due to unequal crossing over between misaligned homologous regions, in fact recombination between polymorphic markers in proximal and distal position referred to the deleted region has been described [37].

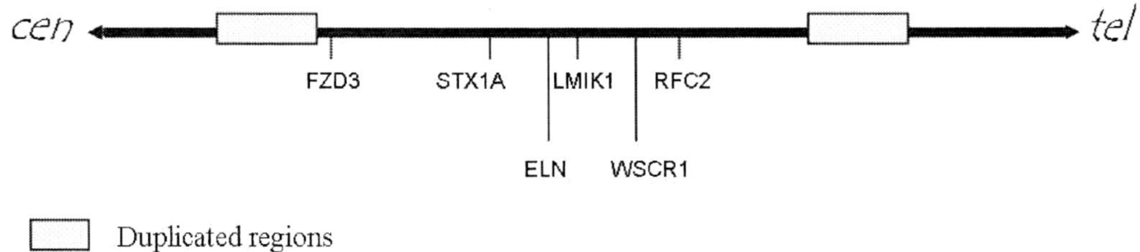

Duplicated regions

Figure 15. Chromosome 7q11.23 trait and duplicated regions.

The regions flanking the deleted sequence have been identified as duplicons. They are about 400 kb long and contain transcribed genes, conserved pseudogenes almost identical, pseudogenes deriving from ancestral progenitors located at other sites on the same chromosome and putative telomere-associated repeats. The repeats contained within the duplicons share about 95% of identity and occur in the same or opposite orientation [38].

Two duplicated regions flanking the deleted interval were been described: they are about 30 kb long and are located in the 3'-region of a transcribed gene proximal to the telomeric region, called *GTF2I*, and a pseudogene proximal to the centromeric region, called *GTF2IP1*, respectively.

They are supposed to be duplicated genes; the telomeric copy seems to be transcribed and alternatively spliced in some different mature mRNA encoding different protein isoforms identical to BAP-135, SPIN and TFII-I, while the centromeric copy is transcribed into a small mRNA probably not translated: it could be a pseudogene.

The recombination mechanism of unequal crossing over between chromosomes or chromatids paired with *GTF2I* and *GTF2IP1* causes the loss of the active copy (*GTF2I*) in all WBS subjects with a deletion of the elastin gene; so they are hemizygous for that gene and produce only half-normal dosage of its products that have been predicted to be multifunctional members of an eukaryotic transcription factor complex. BAP-135 has been found to be a target of Bkt phosphorylation during B cell receptor activating pathway; actually, B cell function seems to be normal in WBS patients, so it has been speculated that, as BAP-135 is expressed in many foetal tissues, its role extends beyond signalling in B cells. It can be a target for tyrosine kinases other than Bkt (Bkt expression is B cell specific) and it can interact not only in the cytoplasm as with Bkt, but also have a nuclear function as GTF2I/BAP-135 sequence includes putative nuclear localization signals.

Moreover, a weak similarity between a region of GTF2I/BAP-135 and some vegetal and bacterial transposases has been found. It could act regulating the recombination events that affect just the 7q11.23 chromosomal region as a trasposon-like element has been identified in the duplicated sequence; so GTF2I/BAP-135 could have a role in the meiotic recombination characteristic of WBS, acting in breakage, exchange and rejoining of DNA segments in the previously mentioned genetic position, leading to the deletion [37].

Other identified products of the deleted region that could affect subjects with WBS, are SPIN, a multifunctional DNA binding protein, that binds to *c-fos* promoter and interacts with the homeobox protein Phox1, and TFII-I that is an essential component of a transcription factor complex. They correspond to the shortest of the different splice variants of *GTF2I* [39].

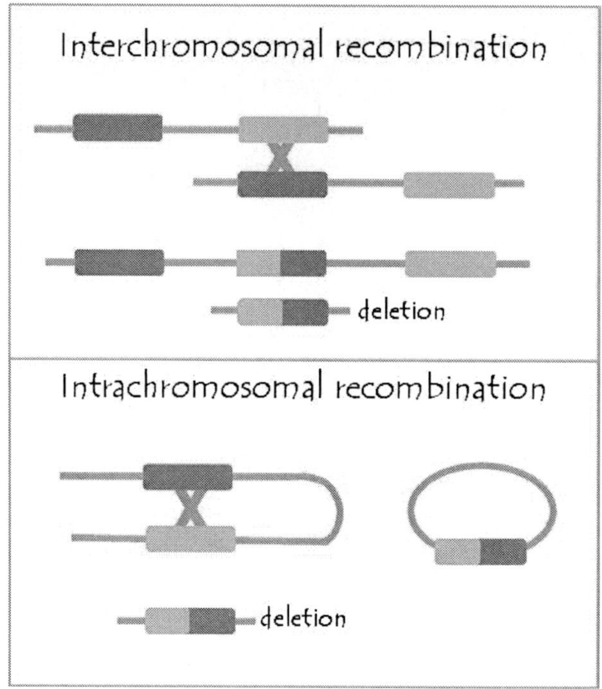

Figure 16. Chromosomal rearrangements in WBS.

The cardiovascular aspects of the disorder are supposed to be caused by haploinsufficiency for the elastin gene (located in the deleted region) because patients with isolated elastin deficiency show only this one of the WBS clinical features. This supposition has been confirmed by the finding of a WBS family (five members) with a balanced translocation t(7;16)(q11.23;q13) in which the breakpoint falls within the elastin gene, which is then disrupted. Although the patients present the same disruption of the interested gene, clinical features of the affected subjects range from full WBS to minimal signs, reflecting a possible variegation effect.

The breakpoint position has been identified within the fifth intron of the elastin gene (on chromosome 7) and the first intron of the TM7XN1 gene (on chromosome 16) in the same manner in all mutation carriers. As isolated mutations of ELN gene is associated just with supravalvular aortic stenosis, but not with the other clinical signs of the syndrome, the complete set of symptoms which characterize WBS can not be explained only with the interruption of the ELN coding gene, but a further genetic mechanism must be involved in the disease development.

The transposed sequence resulting from the genetic variation is characterized by continuous ORFs between elastin and *TM7XN1*, so it has been proposed a possible role of the fused protein. Actually, *TM7XN1* protein product (GPR56) role has not been defined yet, despite it shows an identity of 32% with a member of a subclass of the class B secretin-like G-protein-coupled receptors. Anyway, it is not likely that the new protein retains elastin physiologic role, either because of the complex tridimensional elastin structure and the presence of supravalvular aortic stenosis in carriers of the genetic variation.

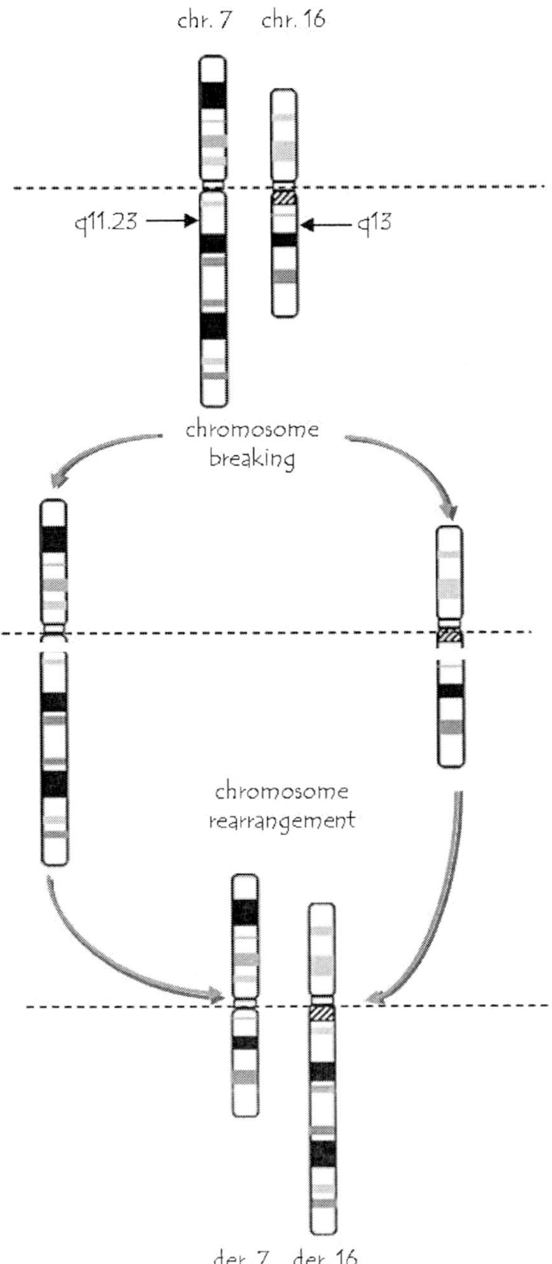

Figure 17. Chromosomes 7 and 16 and their rearrangements in WBS.

It is likely that a positional effect is causative of the phenotypical differences between the affected members, although unidentified genetic differences, environmental or stochastic factors can contribute to the disease expression [36].

The positional effect has been described for mutations that alter gene expression through long-range influence on chromatin structure or variations in distal regulatory elements of a gene. As the definition implies, the causative mutation can be located also several hundred kilobases away from the affected gene [40].

The positional effect is essentially caused by the interaction of different factors in determining correct gene expression:

- promoter region, the site where the DNA transcription machinery recognize the proximal gene;
- enhancer/silencer elements, genomic regions containing binding sites for transcription factors; they participate in assembling the transcription complex at the promoter site independently from the distance and the orientation with respect to the promoter;
- surrounding chromatin structure, it must allow the transcriptional proteins to bind the genetic region; if the chromatin environment is not permissive it isn't possible gene transcription.

The translocation of a gene from one to another genomic site can lead to variations in the genetic environment introducing next to the translocated gene new regulatory elements such as the enhancers/silencers, promoters or in a more condensed chromatin region [41].

The positional variegation effect is a consequence of the positional effect: when a gene translocates from a locus to another one, if its position is near to a heterochromatin/eucromatin boundary, it is possible that a different pattern of expression characterizes different cells determining a diagnostic mosaic phenotype [42].

As previously mentioned, duplicons contain repeated inverted sequences that can cause misalignment and unequal crossing over leading to inversion of the included segment. This has been confirmed through the finding of two atypical WBS patients that don't show the usual deletion, but a clear paracentric inversion of the 7q11.23 region and two typical WBS patients. Moreover, also one parent of an atypical WBS subject (which showed the inversion) and one parent of a typical WBS parent (with a balanced translocation), both phenotypically normal, have been found to have the invertion, suggesting that this variant can likely lead to chromosomal rearrangements in future generations.

Finally, most of the inverted carriers don't show obvious phenotypic features, but at least in two individuals the inversion is associated with several WBS syndrome [38].

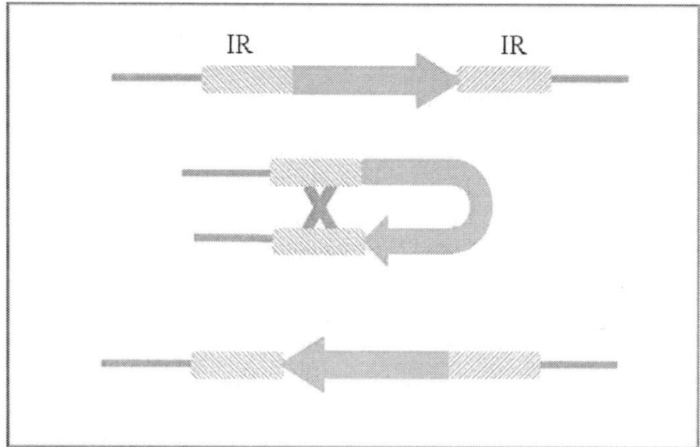

Figure 18. Inversion mechanism.

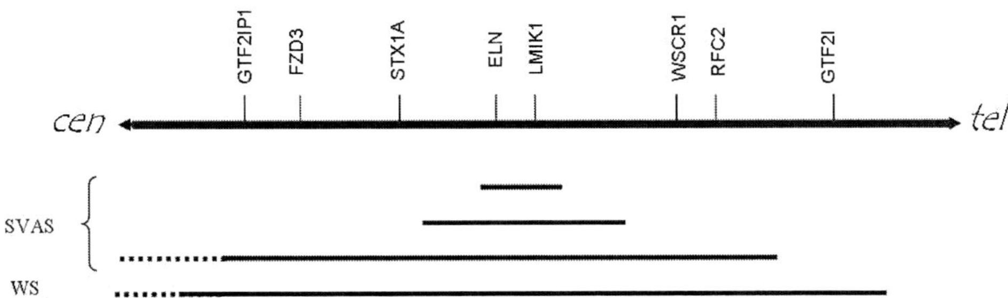

Figure 19. Different described deletions and related phenotypes in WBS.

Another gene that has been identified in the deleted region is *GFT3* (it covers the 10% of the complete deleted region); it results deleted in patients with full WBS but not in subjects with smaller deletions and characterized only by SVAS and not other clinical features, suggesting a principal role in developing the WBS phenotype.

It has been found expressed in adult skeletal muscle and heart. The protein product of that gene has a tridimensional structure similar to transcription factors and an aminoacidic sequence homologous to TFII-I (especially in the N-terminus region, 66 identical aminoacids, and a stretch of 90 aminoacids that fold into a helix-loop-helix repeated motif, present in six copies in TFII-I and five copies in GTF3). TFII-I (encoded by *GTF2I* gene, belonging to the same deleted genome region) is known to bind both DNA (Inr, Inr-like and E-box elements) and proteins (serum-response factor, Phoxamd USF1), so, on the basis of that homology, GTF3 is predicted to have a similar role. Moreover, a high level of similarity is given by the comparing with the sequence of a protein involved in regulation of skeletal muscle gene expression, MusTRD1; it has an identical sequence, except at two positions, to a part of GTF3 protein: a region of both sequences is almost completely overlapping. As MusTRD1 binds an enhancer region upstream to the gene of troponin I, a similar function is predicted also for GTF3; this hypothesis is supported by the expression profile previously described. It is a likely function that links the protein reduction in heterozygosity caused by the deletion of the 7q11.23 region and some symptoms of WBS: many young and old affected subjects appear to suffer from muscle problems, ranging from simple cramps to severe pain during long-time walkings [43].

Genetic Test

Because of the variable clinical features, the genetic test is suggested every time a patient shows supravalvulary aortic or pulmonary stenosis. An early diagnosis for Williams-Beuren syndrome allows, as well as appropriate testing for other members of the family, also an intervention to help affected children in learning [2].

Genetic tests and prenatal diagnosis are performed using FISH-metaphase analysis or targeted mutation analysis (by real-time quantitative PCR to determine the dosage of *ELN*, *LIMK1* and *GTF2I* genes, genomic microarray analysis CGH, detection of heterozygosity by testing Short Tandem Repeats); over 99% of individuals with clinical diagnosis of Williams-Beuren syndrome have 7q11.23 deletion.

Prenatal testing is suggested to unaffected parents who have had a child with WBS: it is necessary to avoid the recurrence risk associated with the possibility of germline mosaicism or inversion polymorphism (see the following) [25].

Genetic Counseling

The deletion characteristic of WBS is inherited in an autosomal dominant manner.

Genetic counseling is intended to assess the risk of affection for family members; as the mutation mechanism arises *de novo* in most cases, often the parents of a proband are not affected, so genetic testing of their genome is not recommended in absence of clinical features. Anyway, some studies showed an inversion of the interested chromosomal region in unaffected parents in 25-30% of cases, but no diagnostic testing in available for this situation, yet.

The risk of affection for siblings of a proband are dependent on the genetics of the parents: if they carry the deletion, the inheriting risk is 50%, but if they don't the risk is extremely low as only few cases of familiar affection are reported. For the same reason, offspring of a proband has a 50% of risk of inheriting the deletion.

In affected individuals, prenatal genetic testing is recommended [25].

Alagille Syndrome

Alagille syndrome is characterized by bile duct paucity in association with at least three of the following clinical signs: cholestasis, cardiovascular or skeletal or ocular anomalies, typical facies.

Cardiovascular defects are present in most affected patients (80-100%) with a prevalence of peripheral pulmonary hypoplasia, Tetralogy of Fallot (the most common complex cardiac defect, present in 7-16% of affected people [25]), septal defects, pulmonary valve stenosis, [44] truncus arteriosus, secundum atrial septal defect, patent ductus arteriosus and ventriculoseptal defect [45]. Alagille syndrome is an autosomal dominant condition [18] characterized by low penetrance and great variability of expression [44].

Genetics of Alagille Syndrome

There isn't a common identified genetic variation for all subjects, but in a minority of them a deletion at chromosome 20p12 has been found; comparison of the cytogenetic breakpoints in affected patients has confirmed the assignment of the causing gene to 20p11.23-12 chromosomal region [44]. The first identified subject with a genetic variation carried a deletion of almost the entire short arm of chromosome 20, but other affected patients have been demonstred to have various smaller deletions [46].

It was previously proposed that Alagille syndrome could be caused by gene-continuous deletion belonging to the interested region; but large deletions, detectable by cytogenetic methods is noted only in few patients, while sometimes the deletion interests such a small segment that it must be investigated by molecular analysis. Despite to the gene-continuous hypothesis, further than the deletion, a balanced translocation involving chromosome 20 and 2 has been found in a family with a clear Alagille syndrome patient and two other family members showing mild phenotypical features.

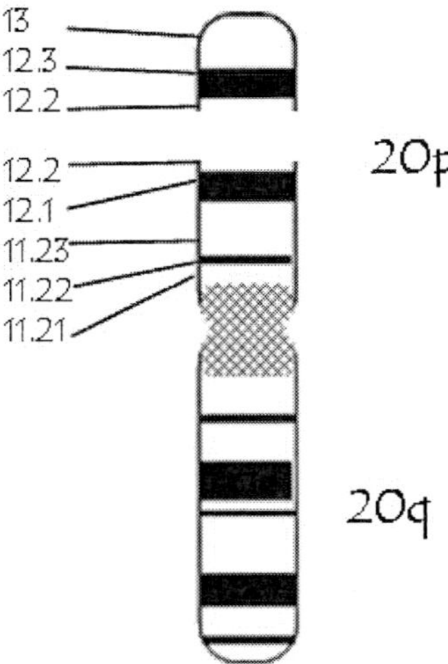

Figure 20. Breaking point on chromosome 20 in AGS.

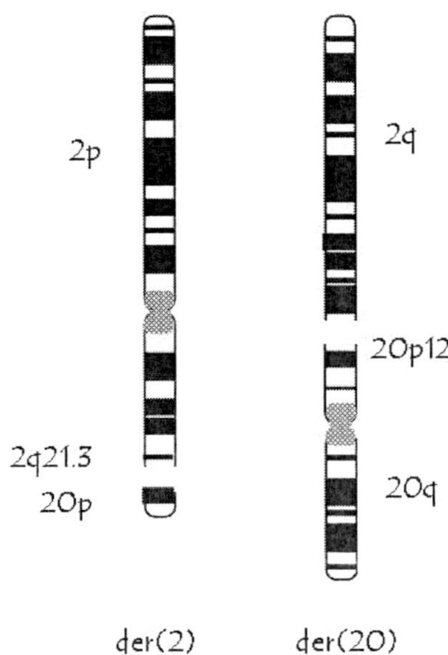

Figure 21. Translocation between chromosomes 2 and 20 in AGS.

So, at least in that family, the balanced translocation associated to the disease suggests a mechanism of interruption of a gene determined by the breakpoint or an alteration of the expression rate of the gene caused by a positional effect after the translocation. A third hypothesis can be a submicroscopic deletion (involving one or more genes) at the breakpoint (it has been validated for other genetic diseases as aniridia) [44].

The deleted region has been investigated in patients with small deletions to identify the possible candidate gene for the expressed phenotype [46].

That region contains the *JAG1* gene, encoding for a Notch ligand protein product, which is mutated in 90% of subjects with Alagille syndrome and mutation analysis for this disease is now available [35] (by FISH analysis). 5-7% of cases are characterized by deletions of *JAG1* gene [18]. *JAG1* product is a conserved trasmembrane protein; it contains several evolutionary conserved domains, such as the signal peptide, conserved DSL domain for the interaction with the receptor, 16 EGF-like repeats, a Cys-rich tract and a transmembrane region involved in Notch signalling.

That pathway plays an important role in cell differentiation and in cell fate determination. Other than the subjects with the deletions, several Alagille patients have been shown to have JAG1 mutations; more than 200 different mutations have been described [47].

JAG1 has been shown to be expressed during the development of mammalian heart valves as well as the great vessels with particular abundance in the outflow tracts, pulmonary artery, ductus arteriosus and descending aorta according to the cardiac diseases of Alagille syndrome [45]. The same has been found in human embryos, in which *JAG1* has been demonstrated to be expressed also in ciliary body region, portal vein, hepatic artery, ductal plate, biliary epithelium, major arteries, meso- and metanephros; the embryonic expression of *JAG1* correlates with many of the tissues affected by Alagille syndrome [48]. A part of the identified variations maps to the extracellular domain; these mutations are very heterogeneous, indeed some cause premature protein truncation, some aminoacidic substitution and others mRNA splicing variants. The highest rate of mutation is located in the gene region encoding for the conserved domains: 30% in the DLS domain, but also in EGF-like repeats and in the Cys- rich sequence [47].

Not only the truncated protein (including small nucleotide insertion/deletions and nonsense mutations) [49], but also the mutated protein is predicted to be impaired in its function; some missense mutation changes an Arg aminoacidic residue that may play an important role in protein folding (both in determinating the secondary and the third structure), so the result can be haploinsufficiency or a dominant negative effect [50].

Some studies have investigated the possible molecular mechanism that, from the genetic mutation of *JAG1* gene, leads to the expressed phenotype. *JAG1* is expressed during foetal development in various tissues involved in Alagille syndrome (among them there are spleen, aorta, kidney, lung, skeletal muscle and liver, they are ordered by highest expression level). The presence of all categories has also been identified of mutant transcripts, in those tissues; a comparison has been done between the different kinds of mutation.

Missense transcripts have a relative stability and can be translated into mutated proteins (it has been confirmed by *in vitro* translation) that are not able to reach and be retained in the endoplasmic reticulum to be processed; some found variations affect the DLS or EGF-like motif by introducing odd numbers of cysteine residues and leading to the 0disruption of the correct disulfide bonds. Haploinsufficiency is the most likely cause of the correspondent phenotype, but a dominant negative effect can not be excluded [49]. Some missense mutations at the 5'- end of the gene have been studied for their influence on protein function: JAG1 –R184H and –L37S are improperly targeted through the cell (they have a different pattern of glycosilation if compared with the wild type protein) and fail to reach cell surface (the lost their function). These mutations determine a conformational change that inhibits targeting to the Golgi apparatus for proper processing; indeed mutant protein is retained in the cytoplasm and its export from the ER is prevented. ER possesses a mechanism that prevents the transport of mutant, misfolded or improperly complexed proteins [51].

In-frame deletions cause the loss of a portion of the protein without changing the remaining downstream sequence. Deletions in the N-terminal and C-terminal domain have been identified: the first one is not very conserved, but has been proposed to play a role in the interaction with Notch receptors; the second is affected in the C-terminal region of the Cys-rich domain, without affecting the sequence of the canonical cysteines.

Premature termination codons containing transcripts are submitted to degradation by nonsense-mediated decay; this process doesn't eliminate completely the mutant mRNA and they can be translated into truncated proteins. These proteins are not functional and can cause haploinsufficiency, but a possibility is also that they act in a dominant negative fashion antagonising Notch receptors [49].

The phenotypes of patients with mutations in *JAG1* gene is indistinguishable from the patients carrying the deletion of the chromosome region 20p12 (it contains the entire *JAG1* gene). So, a haploinsufficiency hypothesis has replaced the early gene-contiguous theory. A 5bp GAAAG repeat has been found in exon 16 of the gene by molecular genetic studies; it is likely that this repeated sequence can lead to recombination errors causing the common deletion [50].

Other Alagille syndrome affected subjects, lacking *JAG1* variations, have been found to have mutations in the NOTCH2 receptor [6].

As mild and severe phenotypic features have been identified in carriers of mutated *JAG1* gene, a role of modifier genes has been supposed [52]. *JAG1* mutation (affecting a splice site), both affected by Alagille syndrome, but showing different clinical features (one had more severe cardiac disease, the other accentuated hepatic diseases). A nongenetic theory explains this situation as the result of twinning process, while the genetic hypothesis suggests genetic modifier genes, a second-hit process (the same used to explain some turmors) or mosaicism. The possibility that mutations arising *de novo* in modifier genes is very low, but can not be excluded; the last one hypothesis has been excluded in the specific case because the twins didn't show genetic heterogeneity. Another explanation could be the different amount of functional protein generated in various tissues: the exact protein level, in a certain tissue at a certain time, may depend on the expression rate of each allele and on the splicing efficiency [53].

Another cause for the variability in the phenotypic expression of Alagille syndrome could be also due to the size of the deleted region: in some known disease a largest deletion corresponds to a more severe clinical setting. Alternatively, a variation in expressivity of the causative gene can be supposed [44].

Moreover, genetic mosaicism has been supposed for the different clinical settings of Alagille syndrome affected patients. A case of an affected patients with total *JAG1* deletion that has inherited the mutation by the phenotypically normal mother has been described. The mother was a genetic mosaic for the deletion encompassing *JAG1* gene (9/20 of the peripheral blood cells showed the deletion); presumably the mosaicism involved also the germinal line, and it has been passed on to her daughter. The possibility of such a process must be considered when unaffected parents have an affected child: not only *de novo* mutations, but also parent mosaicism can be supposed. It is of particular importance for the risk assessment of having future affected children [54].

Another cause of the phenotypic variation could be the genetic heterogeneity due to different gene mutations. Indeed, about 90% of Alagille patients have *JAG1* mutations, but the remaining 10% don't have an identified genetic cause yet. Some studies have investigated the presence of mutations in a second gene involved in the same signalling pathway (*NOTCH2*) in patients negative for *JAG1* mutations. Two *NOTCH2* mutations have been identified: a mutation of the splice acceptor site of the exon 33 (c.5930-1G>A) and in the coding sequence (c.1331G>A). The first one determines the formation of a splicing product with a premature termination codon within exon 34, but the transcript doesn't undergo nonsense mediated decay because it is in the last exon. The result is a protein lacking three of the seven ankyrin repeats (they are necessary for protein-protein interactions; Notch interacts with nuclear cofactors through these sequences). The second mutation changes the aminoacidic sequence replacing a tyrosine residue with a cysteine in the 11th EGF-like repeat (C444Y). The EGF-like motif is composed of six Cys residues and forms three intramolecular disulfide bonds; they are essential for protein folding. So the gain or loss of any Cys invovlves a change in the three-dimensional structure.

Other than the consequences of the mutated protein, there could be some *JAG1* or other *NOTCH2* polymorphisms that influence the final (variable) phenotype [55].

Genetic Test

Genetic testing and prenatal diagnosis (on foetal cells derived by amniocentesis or chorionic villi) is performed to identify a mutation in the interested genes (*JAG1* and *NOTCH2*) with different methods: analysis of the entire coding region by sequencing analysis or mutation scanning, sequence analysis of selected exons, linkage analysis, FISH-metaphase, deletion/duplication analysis.

In patients with clinical diagnosis of the syndrome, genetic test is confirmative of the disease. Sequence analysis of *JAG1* gene shows mutations in about 88% of patients, while FISH analysis detects a micordeletion of 20p12 chromosomal region only in 7% of subjects (if a deletion is found, whole chromosomal analysis can detect possible translocations or inversions). *JAG1* analysis are performed expecially on exons 1-6, 9, 12, 17, 20, 23, 24 (they contain about the two/thirds of the mutations). With a lower frequency there are mutations in *NOTCH2*: fewer than 1% of cases with Alagille syndrome; it is usually investigated in patients with AGS and no *JAG1* mutation or deletion but no laboratories perform prenatal genetic analysis on this gene.

Genetic tests assess the presence or absence of a *JAG1* mutation or deletion, but can not predict the severity of clinical features [25].

Genetic Counseling

Alagille syndrome is inherited in an autosomal dominant fashion, but only the 30-50% of patients have an inherited mutation; it is very frequent *de novo* mutation (about 50-70% of cases).

Genetic counseling is suggested to adults surely affected or at risk of affection, especially before pregnancy.

Parents of an affected proband have some increased risk of transmission to other offspring because of the possibility of undiscovered germline mosaicism, they are affected only in 30-50% of cases (about 50-70% of mutations are due to *de novo* mechanism). Genetic testing of the parents is recommended if they present any clinical feature of the disease, if the proband has an identifiable disease-causing mutation in *JAG1* gene or a microdeletion of 20p12 chromosomal region.

The evaluation of the risk of affected siblings is dependent on parents' genetic condition (if one of the parents is affected, the probability is of 50%, but if no one of them is a carrier, the probability is very low, unless it is a mosiacism condition)

As reported for other autosomal dominant diseases, the risk of transmission to offspring of the proband is 50% and prenatal genetic testing is possible in this situation (both if the parent has a *JAG1* or *NOTCH2* mutation or a FISH-identified deletion). A issue that must be considered is that the finding of a mutation or deletion in the foetus is not predictive of the severity of the disease.

Proband genetic analysis is possible in the cases of unclear clinical signs (suggesting the possibility of AGS) [25].

Noonan Syndrome

Noonan syndrome is an autosomal dominant condition with high expression variability; about half of the cases are due to *de novo* mutations [18].

Noonan syndrome (NS) is characterized by various features; among them we found short stature, particular facies, webbed neck, chest deformity and cardiovascular diseases. Most of Noonan syndrome patients have cardiac defects (80-90%) with a prevalence of valvar pulmonic stenosis (50-62%) and hypertrophic cardiomyopathy (20%), but also with secundum atrial septal defect, atrioventricular septal defect, mitral valve abnormalities, aortic coartation and Tetralogy of Fallot [56].

The incidence of NS is reported to be between 1 in 1000 and 1 in 2500 live births [57].

Genetics of Noonan Syndrome

Some genes have been associated with this syndrome, such as *PTPN11* (encoding a protein tyrosine phosphatase called SHP-2, involved also in semilunar valve formation mutated in about 50% of cases), *SOS1* and *KRAS*, so defining NS as a genetically heterogeneous disease inherited in an autosomal dominant fashion [58]. SHP-2 is a member of a subfamily of protein tyrosine phosphatises (PTPs). Its structure is characterized by two amino-terminal src-homology (SH2) domains arranged in tandem (N-SH2 and C-SH2), a PTP and a carboxy-terminal tail. SHP-2 has an ubiquitous expression and is involved in mesodermal patterning, semilunar valvulogenesis and hematopoietic cell differentiation. Its pathways include the cell response to growth factors, hormones, cytokines and cell adhesion molecules, indeed it is necessary for the activation of Ras/mitogen-activated protein (MAP) kinase.

NS causing *PTNP11* mutations, which involve SHP-2 domain are predicted to determine gain of function due to stabilization of SHP-2 domain in the active conformation.

The identified mutations cause the change in the correspondent aminoacidic residue, especially in the N-SH2 and PTP functional domain [56]. Indeed, most of the mutations found in NS patients resides in and around the N-SH2/PTP interaction surface and are supposed to impair N-SH2/PTP interaction leading to an increased phosphatise activity. The only deletion found in *PTPN11* is del60Gly; the correspondent molecular situation is also an excessive phoshatase activity, because Gly 60 is a fundamental amino acid in N-SH2/PTP interaction.

Another possible mechanism is the acquired function in promoting the dissociation between the two domains or determine a conformational change impairing their interaction: this has been proposed for Thr42Ala change because Threonine 42 is out of the interactive surface.

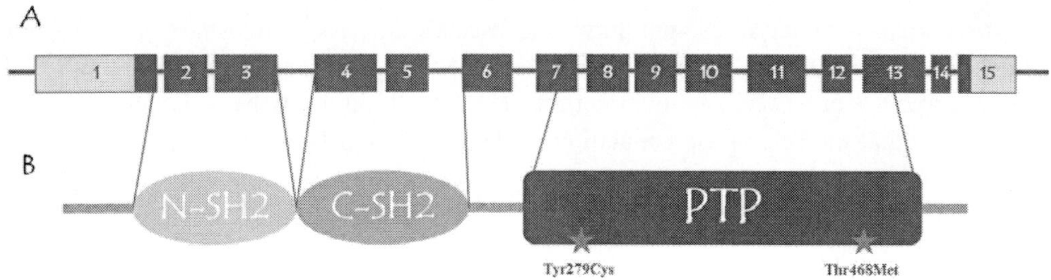

Figure 22. PTPN11 gene (A) and protein (B) structures; in PTP domain two mutational hotspots are indicated.

Some mutations have been associated to specific phenotypic features. Some examples are Y62D and T31I, identified in NS patients without juvenile myelomonocytic leukemia (JMML) but then found in NS patients with JMML. Controversially, S502T mutation has been firstly described in NS patients with JMML and then found in the other ones; this findings are in accord with the supposed heterogeneity of mutations associated to the disease. Another particular mutation is Y279C: it has been found in one NS patient and then identified in several LEOPARD affected subjects [59].

Most of the patients with *PTPN11* mutations are affected by pulmonary valve stenosis while those without *PTPN11* mutations show a higher incidence of hypertrophic cardiomyopathy. The most frequent mutation is Asn308Asp, found in the 25% of the NS patients with a mutation in PTPN11. It is often found in familial cases but it is not associated with any developmental abnormality. The mutation Thr73Ile, instead, has been associated with a predisposition in myeloproliferative disorder [57].

Genetic heterogeneity allows to use genetic testing for *PTPN11* mutations as a confirm for the diagnosis of NS, but the lack of genetic variations isn't an exclusion for the syndrome. Moreover, *PTPN11* mutations are indicative of the need of a genetic counselling for the carrier subject and some differences between mutation-positive and mutation-negative affected patients can be revealed during the follow-up of the subjects [59].

The syndrome often arises by *de novo* mutations of paternal origin [57].

Because of the genetic heterogeneity, although there are now available genetic tests for detecting mutations in that genes, they can only confirm but not exclude the diagnosis of NS; moreover, genetic tests are useful also for revealing mutations in foetuses from known affected parents [2]. Gentic analysis is performed by different techniques to investigate *PTPN11*, *SOS1*, *KRAS* and *RAF1* genes: analysis of the entire coding region by sequencing analysis or mutation scanning, sequence analysis of selected exons, linkage analysis, FISH-metaphase, deletion/duplication analysis [25].

The frequency of congenital heart diseases in Noonan's syndrome in about 50-90%, with a prevalence of stenotic pulmonary valve (20-50% of cases), dysplastic valve, and hypertrophic cardiomyopathy (20-30% of cases). Other frequent cardiac defects are ASD, VSD, sopravalvular pulmonic stenosis, branch pulmonary artery stenosis, TOF, coarctation of the aorta and partial AVSDs [18].

Genetic counselling is indicated before the child is born to explain the mechanisms of occurrence of NS in the foetus and the recurrence risk in the family, to suggest any other study to do to confirm the diagnosis, to suggest currently available treatments and interventions help the management of the affected child. Indeed, NS must be supposed in foetus with polyhydramnion, pleural effusions, oedema and increased nuchal fluid, also if a normal karyotype is found. If there is a diagnosis of NS or a first-degree parent is affected by the syndrome, some additional exams are requested (obstetric ultrasound at 12-14 and 20 week's gestation; foetal echocardiography is indicated at 18-20 weeks' gestation). Also parents' physical examination for features of NS can be done and DNA testing of the blood, chorionic villi or amniotic fluid can be conducted [57].

NS is related to three other clinical conditions due to *PTPN11* variations: LEOPARD syndrome (typical features are multiple lentigines, electrocardiographic conductions anomalies, ocular hypertelorism, pulmonic stenosis, abnormal genitalia, retardation of growth and sensorineural deafness), Noonan-like (presents the same clinical signs of NS plus giant cell lesions of bones) and hematopoietic disorders [60].

In LEOPARD syndrome facial dysmorhism includes ocular hypertelorism, palpebral ptosis and low set ears; stature under the 25th percentile and cardiac defects are mostly represented by hypertrophic cardiomyopathy of the left ventricle and ECG anomalies [61, 62]. Clinical diagnosis may be suspected when multiple lentigines and two other features are found; alternatively, if no multiple lentigines are present, three other disease features and an affected close relative confirm the diagnosis [61].

Approximately 85% of affected individuals have heart defects, with a frequency of more then 70% of hypertrophic cardiomyopathy (it usually appears during infancy), about 25% of cases have pulmonary valve stenosis and a minority of persons with abnormalities of the aortic and mitral valve [25].

Some studies show the prevalence of left ventricular hypertrophy in population affected by LS and the lower frequency of right ventricular hypertrophy; single patients have other cardiac abnormalities such as left ventricular apical aneurysm, left ventricular noncompaction, isolated left ventricular dilatation and atrioventricular canal defect [63].

Long term prognosis seems benign in LS patients with mild cardiac abnormalities, while patients affected by LVH may develop significant symptoms and arrhythmias [62].

LS is characterized by genetic heterogeneity, too. It has been confirmed through linkage analysis: about he 85% of LEOPARD patients have *PTPN11* heterozigous missense mutations in exons 7, 12 or 13; recently missense mutations in *RAF1* gene in few *PTPN11*-negative patients. It is an autosomal dominant fully penetrant disease and may be sporadic or inherited. *RAF1* is a threonine-serine protein kinase involved in *RAS* pathway (it acts downstream) and is one of the three mammalian isoforms (*RAF1* mutations also characterize a subset of NS patients, mostly with HCM). The 5% of LS patients don't have neither *PTPN11* nor *RAF1* mutations, suggesting other possible causative-genes.

Some associations have been done between mutations in exon 7 and 12 of *PTPN11* and HCM and between exon 8 an PVS. Mutations in exon 13 are often associated with rapidly progressive severe biventricular obstructive HCM. Specific association studies have been conducted to investigate some *PTPN11* mutations: mutation of the Thr468 aminoacidic residue is associated with short stature and Tyr279 with deafness. However, the genetic difference between NS and LS resides in the effect of the mutations: gain of function in NS and reduced protein tyrosine phosphatase activity in LS [61].

Association studies have investigated the influence of *PTPN11* mutation on the severity of LVH (the most common defect in patients with LS): no association between a *PTPN11*-positive or –negative genotype and the magnitude of LVH was found. However, patients without *PTPN11* mutations carry an elevated risk of arrhythmic events, based on the increased left ventricular dimensions, but the genotype seems not directly related to fatal events in patients with LS (fatal events have been found in the study population in 50% of cases independently by the presence or absence of *PTPN11* mutation) [62].

Some other association study showed that subjects carrying *RAF1* Leu613Val and Ser 257Leu substitutions have a full LS phenotype with multiple lentigines, CLS and HCM. Leu613Val mutation (in patients with HCM) is responsible of an increased kinase activity and enhanced ERK activation, supporting the role of RAS signalling in cardiomyocyte hypertrophy pathogenesis [61].

The clinical features of NS and LEOPARD syndrome overlap, but some differences are found in the skin pigmentary: NS is characterized by pigmented nevi and cafè au lait spots, while LEOPARD syndrome manifests cafè au lait spots during infancy and multiple lentigines after 5-6 years of age.

Genetic findings allowed to strongly connect the two diseases: *PTPN11* was found mutated in some LEOPARD patients. Missense mutations in exon 7 (836A>G; Tyr279Cys) and exon 12 (1403C>T; Thr468Met) can affect protein activity because they fall in the regions encoding for the phosphotyrosine phosphatase (PTP) domain, relating the phenotypic feature of skin pigmentation of the two syndromes. Both Tyr279 and Thr468 are highly conserved in tyrosine phosphatise enzymes, suggesting an important role of the two aminoacids in the PTP domain. The Thr468 plays a critical role in Tyr-phosphatase activity (belongs to the tyrosine-specific protein phosphatase's active site) and is involved in several hydrogen bonds toward the solvent; this last interaction contributes to the stability of the protein structure. However, due to different mutation findings, genetic heterogeneity is supposed [64]. Mutations in exon 13 seems to have an high specific and positive predictive accuracy for adverse events in patients with LVH and LS [62].

Genetic Test

Clinical testing to assess the genetics of the Noonan Syndrome affected subject are performed by sequence analysis of all exons of *PTPN11* (about the 50% of the cases show missense mutations in this gene); all exons of *KRAS* (less than the 5% have mutated *KRAS*); exons 1-23 of *SOS1* and exons 1-17 of *RAF1*. FISH analysis for the detection of deleted *PTPN11* show positive result only in about 1% of cases (deletions in the gene are very rare).

Genetic tests to identify the presence of a mutation in one of the causative genes for Leopard syndrome are performed by sequence analysis of the entire coding region or selected exons; alternatively an indirect mutation scanning allows the analysis of the entire coding region.

The usual genes that are investigated to diagnose Noonan syndrome are *PTPN11* and *RAF1*. *PTPN11* has been found mutated in 90% of individuals affected by NS, while *RAF1* in about 3% (analysed exons 7, 14 and 17) [25].

Genetic testing is a useful tool to confirm the clinical suspicions; particularly, in the first month of life phenotypic features are not well-defined (an example are the lentigines) and appear later. Clinical studies followed by molecular testing show the confirmation of the diagnosis in most of the cases, but a little percent of the subjects may have a different (clinically similar) disease (as neurofibromatosis type 1) [65].

Genetic Counseling

Noonan and Leopard syndrome are inherited in an autosomal dominant manner.

Most patients have *de novo* mutations and only some cases are familiar.

Parents of affected subjects are investigated only after a clinical evaluation for symptoms of the syndrome, apparently healthy parents can show genetic mutations if the signs of the syndrome are only mild (for Noonan syndrome, an accurate evaluation of the phenotypic features at various ages is recommended). If parents don't show genetic mutations of one of the causative genes, a possible *de novo* mutation or germline mosaicism could be a reason of the development of the disease; eventual non-medical explanation is the possibility of alternate paternity or maternity due to assisted reproduction.

The assessment of the risk of affection for siblings of the proband and other family members is dependent on the genetic status of the parents (50% of probability if a parent is affected and very low if parents are unaffected), while offspring of the proband has a 50% of chance to inherit the mutated gene.

The optimal time to have a genetic counseling for probands is before pregnancy to assess the genetic risk and the planning of prenatal testing of the foetus [25]. Prenatal diagnosis between NS and LS isn't easy to perform because of the overlapping clinical features.

Some different examination plans are indicated for high- and low-risk pregnancies. For the first instance molecular genetic testing after amniocentesis or chorionic villi sampling and ultrasound examination for clinical signs are performed. For low-risk pregnancy, ultrasonographic exams are enough to exclude the possibility of the disease.

LS should be supposed in foetus with normal karyotype and HCM; if it is suspected, physical examination of the parents is suggested. Then, if one parent results affected, specific analysis of the foetus are performed (obstetric ultrasounds, foetal echocardiography at 20 weeks of gestation, and DNA testing on chorionic villi or amniotic fluid).

Genetic counselling is indicated to revise three generation family tree especially to evaluate the recurrence of skin and cardiac anomalies, short stature and learning difficulties; examination of growth parameters, facial dysmorphism of the propositus and complete clinical and cardiological examination of the parents (inclusive of echocardiogram and ECG if possible); revision of the recurrence risk; raccomandation for clinical and molecular tests (first to evaluate mutations in *PTPN11* and then *RAF1* in *PTPN11*–negative subjects); the management and the follow up of the disease [61].

Holt-Oram Syndrome

Holt-Oram syndrome (HOS) is an autosomal dominant [66] heart-hand disease and often occurs as sporadic cases. The principal clinical signs that allow the diagnosis are congenital heart defects in association with upper-limb deformities due to mutations in *TBX5* gene (about 70% of cases), encoding for a transcription factor just involved in the embryological development of the heart and limb [67].

HOS is estimated to occur in 1/100000 live births and the most common forms of CHD associated with the disease are ASD of the ostium secundum variety and VSD in the muscular trabeculated septum; less frequently, cardiac conduction diseases may also occur [66]. So, the cardinal clinical feature for Holt-Oram syndrome diagnosis is the presence of upper-limb deformation, apart from the cardiac defects [2].

Genetics of Holt-Oram Syndrome

TBX5 was the first single gene mutation described to cause human septation defects [6].

TBX5 is a member of the conserved T-box family of transcription factor genes; other members of this family, are involved in human diseases (*TBX3* mutations cause ulnar-mammary syndrome, *TBX4* is implicated in "small nail patella" syndrome, *TBX22* in X-linked cleft palate with ankyloglossia and *TBX1* in CHD associate with DiGeorge syndrome) [66].

Mutational analysis of the gene have shown that most of the patients carry variations in the coding sequence, but the remaining patients probably have mutations in the regulatory regions of the gene or in the coding region not investigated by the available tests. However, the disease shows allelic heterogeneity with prevalence of frameshift or nonsense mutations that cause haploinsufficiency (less than 50% of the normal protein product is synthesized and it is not enough to have a wild type phenotype), but a minority shows missense mutations that doesn't affect the protein dosage. Furthermore, some patients with duplication of chromosome 12q segment (the genetic region that includes *TBX5* gene) present the same clinical features, although this situation leads to an overexpression of the gene [68, 69]. Only two mutations (T233M and I106V) are missense; thereaminder are predicted to cause haploinsufficiency, according with the supposition that haploinsufficiency of *TBX5* gene product is the main cause of the development of HOS. An interesting correlation between the two missense mutations and heart defects has been identified: mutations in the 5' end of the T-box, significantly affects the heart while mutations in the 3' end of the T-box domain are responsible for upper limb malformations; anyway a population study has found an opposite correlation, with a patient carrying I106V substitution (in the 3' end of T-box) showing upper limb malformation and only non-severe cardiac defects, while a patient with a T233M mutation has ASD and VSD. Other associations between genetic mutations and clinical heart malformations in HOS have been found in some patients: S196ter *TBX5* mutation in a patient with TOF (indicating the possibility of HOS subjects to develop conotruncal malformations); Y136 ter *TBX5* mutation is characteristic of subjects with cardiac conduction problems (one patient had also septation trial defects, but the family study revealed the responsibility of other genes for this malformation); H271ter has been found in a patient with Long QT and HOS (Long QT was not previously associated to HOS, but this patient had no other mutations in LQT1, LQT2, LQT3, LQT5 and LQT6, so *TBX5*, involved in the normal formation of the cardiac conduction system can be causative of the defect).

Moreover, some studies have identified *TBX5* mutations only in patients with clinical features of Holt-Oram syndrome, while some other subjects supposed to have the disease because of few features overlapping with the set of symptoms of the syndrome (they carried also other clinical features not belonging to the syndrome) had no *TBX5* gene mutation.

However, *TBX5* mutations were not identified in all patients with HOS; the reason why it happens can be the limitations of detecting test in identifying mutations in the non-coding or regulatory regions (three splicing variants have been previously found). Another possible explanation can be the incomplete penetrance of the disease [66].

SALL4 gene also has been identified as a candidate gene to cause HOS [25].

Genetic Test

Genetic tests for detecting Holt-Oram syndrome are particularly limited just because of the allelic heterogeneity, so the lack of genetic assessment doesn't mean the exclusion of the positive diagnosis [2].

However, sequence analysis or mutation scanning of the entire coding region or FISH-metaphase analysis are available for diagnostics.

TBX5 gene mutations has been detected in more than 70% of HOS patients. It is investigated by direct sequencing or mutation scanning, deletion or duplication analysis (FISH; it revealed positive results in about 1% of cases).

Usual genetic analysis focuses on the detection of mutations in exons 1-4 of gene *SALL4* (they contain 80% of detected mutations), whereas FISH investigation has a limited utility due to technical limitations [25].

85% of cases result from new mutations [18].

Genetic Counseling

Holt-Oram syndrome is inherited as autosomal dominant characters.

HOS due to *TBX5* mutations is sometimes of familiar origin, but *de novo* cases are the most frequent situations (about 85% of cases).

Most patients with *SALL4* mutations have an affected parent, but are also reported *de novo* mutations (40-50%); however, sometimes, negative family history can hide affected members because of the difficulty in recognizing the set of symptoms of the syndrome. Before investigating parents of a proband, clinical examination must consider phenotypic signs of HOS. As for other genetic inherited diseases, the risk of affected siblings and other member of the family is dependent on the genetics of proband's parents (the possibility of germline mosaicism must be considered in negative parents). The risk of vertical transmission to the offspring is 50% (prenatal diagnosis in affected subject foetus is suggested). Prenatal diagnosis in high-risk pregnancies includes ultrasound examination of the foetus, despite it doesn't exclude the diagnosis of the syndrome (because of the clinical variability). HOS is a variable syndrome, so the effects of modifying genes causing the variable expressivity (found also in individuals having the same mutation) must be considered; equally, the finding of a certain genetic variation is not predictive of the correspondent phenotype [25].

Turner's Syndrome

Turner's syndrome is an aneuploidy syndrome due to the absence of a X chromosome in a female subject (sometimes it is due to a karyotype 46,XX with an abnormal structure of a X chromosome or mosaicism). Phenotypic characteristics are variable among subjects, but commonly there are excess nuchal skin, marked lymphedema (webbed neck, edema of hands and feet), short stature, gonadal dysgenesis, cardiovascular malformations and renal malfomation. Particularly, cardiac malformations are present in 15 to 50% of affected patients with prevalence of bicuspid aortic valve; coarctation of the aorta is present in about 10% of cases, aortic root dilation in 5-10% of cases; also hypoplastic left heart, VSD and conduction defects have been described [18].

Down Syndrome

Down syndrome is due to trisomy of 21 chromosome (particularly the genomic region involved in the development of the disease is 21q22) or to Robertsonian translocation of chromosome 21. Down syndrome is characterized by facial features, mental handicap and cardiac anomalies. Cardiac malformations include VSD (the most common), ASD, patent ductus arteriosus and AVSD [18].

Nonsyndromic Single-Gene Disorders

Some genetic mutations have been found associated to specific CHD not fitted in a syndromic frame. An example is the gene *NKX2.5*, associated to atrial septal defects and atrioventicular conduction delay; another one is the gene *GATA-4*, related to septal defects. These genes have been found mutated in affected subjects and wild type in normal people; they, other than correlate with the enunciated pathologies, are also involved in heart development [2].

7. Genetic Counselling

The definition of the genetic counselling and the correspondent of genetic counsellor has been subject of continuous changes during the past decades. A brief introduction on the definition of these terms follows.

The term "genetic counselling" has been coined in 1947 by Sheldon Reed; he meant that "The primary function of genetic counselling is to provide people with an understanding of the genetic problems in their family". This first definition didn't recognized genetic counselling as a part of the medical treatment, but as a "social work", also because genetics in 1940s was not diffused and well-known as today (laboratory testing for genetic diseases was extremely limited). The role of the genetic counsellor has been better defined in 1950s, when the president of the American Society of human Genetics wrote: "The counsellor must not only be concerned with the specific problem in inheritance risen by a given family but must also attempt to make some assay of the total genetic endowment of the person in question…" and Robert F. Murray Jr (pediatric geneticist at Howard University) added in 1960s: "The physician who counsels must keep in mind the total psychological constellation of the family…". After this first definitions, the figure of the genetic counsellor has been defined as a professional status. Indeed, the attention was focused on the psychological aspect of the counselling, because the counsellor has the role of giving advice to the interested family, a harder work than communicate the assessment of the risk of inheritance as Robert Bringle, an educational psychologist said ("Genetic counselling is… defined as enabling the counselee to comprehend the medical facts of genetic disorders, hereditary risks, and alternatives, as well as to make a healthy adjustment to a family member's disorder and risk of recurrence. The process of learning is broken down into a hierarchical relationship between acquisition, understanding, and personalization of facts and applied to the genetic counselling situation"). So, genetic counselling, more then a medical is view as a psychological encounter in which the geneticist must facilitate clients' ability to use genetic information minimizing psychological distress. However, the development of genetic techniques for discovering the bases of genetic diseases, the introduction of amniocentesis and biochemical analysis to identify congenital metabolism problems in 1970s led to a change in the consideration of the genetic counsellor, focusing on the medical aspect of the counselling; in a workshop organized by the National Genetics Foundation and the National Institute of General Medical Sciences some definitions were given: "Genetic counseling is a communication process which deals with the human problems associated with the occurrence or risk of occurrence of a genetic disorder in a family. The process involves an attempt by one or more appropriately trained persons to help the individual or family to: (1) comprehend the medical facts including diagnosis, probable course of the disorder, and the available management, (2) appreciate the way heredity contributes to the disorder and the risk of recurrence in specified relatives, (3) understand the alternatives for dealing with the risk of recurrence (4) choose a course of action which seems to them appropriate in view of their risk, their family goals, and their ethical and religious standards and act in accordance with that decision, and (5) to make the best possible adjustment to the disorder in an affected family member and/or the risk of recurrence of that disorder". At last, the advances in genomic medicine and in genetic knowledge, requested a final more complex definition of the role of the genetic counsellor; it was approved by the National Society of Genetic Counselor in 2005 and is the following: "Genetic counselling is the process of helping people understand and adapt to the medical, psychological, and familial implications of genetic contributions to disease. This process integrates the following: interpretation of family and medical histories to assess the chance of disease occurrence or recurrence; education about inheritance, testing, management, prevention, resources, and research; counselling to promote informed choices and adaptation to the risk or condition" [70].

A common problem in genetic counselling is the nondisclosure of genetic risk information within families. This is an important issue, because family members could benefit from the knowledge, for example in the choice of reproduction or the management of the risks of the disease. Usually the cause of this phenomenon is the wish of not causing anxiety or alarm among the family, but sometimes the situations can be complicated because of geographical distance, family rifts, divorce, separation or adoption. Moreover, the information provided during a counselling can be misunderstood by the patients, determining a different view also in the risk for other family members. Fortunately, episodes of nondisclosure within family appear to be very rare. Anyway the rate of nondisclosure events arises in the case of some diseases for which there is not an effective therapy, as Huntington's disease to avoid distress in relatives. The role of the geneticist in a similar situation (when is nondisclosure is evident or declared) is to try to persuade client to pass on relevant information at least to members of the family exposed to the risk of inheritance [71].

Another important ethical issue related to genetic counselling is genetic testing in minors. Indeed, as the knowledge of the risk is a personal decision, testing of minors should be decided taking into account the minor's best interests and, if possible, considering the minor's opinion. The influence of the opinion of the child must have a relative role, proportionate to his age and grade of maturity. Most guidelines say that testing of a minor should be postponed to the moment when the child is able to take a reasoned decision, but some of them agree in considering that suggestion must not be valid when exceptional situations are encountered. Even though, maintaining child's autonomy, confidentiality, and privacy are fundamental right of children. This different interpretations are due to the lack of a conclusive evidence that the discovery of being a carrier of a genetic mutation can harm children psychologically. Anyway, also if the test would be postponed, is necessary to advise the interested child about the risk of knowing to be a carrier and must be avoided situations in which children don't know about their genetic risk. And if the final decision is or not to perform the test, genetic counselling should be available to minors for all the life to help to consider undergoing carrier testing or to guide in accepting the possible carrier status [72].

The development in the discovery of the genetic causes of cardiovascular diseases has an important impact on the risk, the treatment and the counselling for affected subjects. Anyway, only few assessed genetic CVDs have a proper genetic test, although continuous researches are aimed to increase the number of CVDs tested by DNA-based methods.

Genetic counselling is a useful tool to inform the family of the affected subject about the hereditary risk and to suggest genetic testing for other members of the family which could eventually benefit from medical treatment. It is based on clinical characteristics of the affected subject, molecular analysis, family history [1].

In paediatric patients the role of the geneticists is focused on the short- and long- term medical and developmental outcomes and on the risk of affection for the parents; in adult subjects, instead, it is also important to point out the risk of having affected child and which could be the appropriated genetic tests to detect eventual transmission to the offspring.

CHDs have been for long time classified as diseases clustered in families. Some of them show a multifactorial fashion of inheritance, while others (only few diseases) have a mendelian inheritance. The last ones have been particularly useful to identify single-gene related diseases, extending the possibility of preventing the development of the disease and to better understand the mechanisms that lead to the disease. Anyway, only few CHDs belong to mendelian group of diseases, suggesting that CHDs often represent complex multifactorial conditions [18].

Appendix I

Text Abbreviations

ANF	Atrial Natriuretic Factor
Chisel	potential downstream gene target or the cardiac homeodomain Nkx2-5
Irx5	Iroquois-related homeobox gene family
SERC2a	encodes the sarcoplsmic reticulum calcium pump, a key component of cardiac excitation-contraction coupling
Tbx5	T-box transcription factor gene
Hand1	bHLH transcription factor gene eHand/Hand1

References

[1]   D.K. Arnett, A.E. Baird, R.A. Barkley, C.T. Basson, E. Boerwinkle, S.K. Ganesh, D.M. Herrington, Y. Hong, C. Jaquish, D.A. McDermott and C.J. O'Donnell. Relevance of

Genetics and Genomics for Prevention and Treatment of Cardiovascular Disease: A Scientific Statement From the American Heart Association Council on Epidemiology and Prevention, the Stroke Council, and the Functional Genomics and Translational Biology Interdisciplinary Working Group *Circulation* 2007;115:2878-2901.

[2] M.E. Pierpont, C.T. Basson, D.W. Benson, Jr, B.D. Gelb, T.M. Giglia, E. Goldmuntz, G. McGee, C.A. Sable, D. Srivastava and C.L. Webb *Genetic Basis for Congenital Heart Defects: Current Knowledge*: A Scientific Statement From the American Heart Association Congenital Cardiac Defects Committee, Council on Cardiovascular Disease in the Young: Endorsed by the American Academy of Pediatrics *Circulation* 2007;115:3015-3038.

[3] Hirschhorn JN, Daly MJ. Genome-wide association studies for common diseases and complex traits. *Nat Rev Genet.* 2005;6:95–108.

[4] K.J. Jenkins, A. Correa, J.A. Feinstein, L. Botto, A.E. Britt, S.R. Daniels, M. Elixson, C.A. Warnes, C.L. Webb Noninherited Risk Factors and Congenital Cardiovascular Defects: Current Knowledge: A Scientific Statement From the American Heart Association Council on Cardiovascular Disease in the Young: Endorsed by the American Academy of Pediatrics *Circulation* 2007;115;2995-3014I.

[5] V.M. Christoffels, P.E.M.H. Habets, D. Franco, M. Campione, F. de Jong, W.H. Lamers, Z.Z. Bao, S. Palmer, C. Biben, R.P. Harvey, A.F.M. Moorman. Chamber *Formation and Morphogenesis in the Developing Mammalian Heart Developmental Biology* 2000; 223, 266–278.

[6] J. Ransom D. Srivastava. *The genetics of cardiac birth defects Seminars in Cell and Developmental Biology* 2007; 18:132-139.

[7] J.S. Kim, S. Virágh, A.F. M. Moorman, R. H. Anderson, W.H. Lamers. Development of the Myocardium of the Atrioventricular Canal and the Vestibular Spine in the Human Heart *Circ. Res.* 2001;88:395-402.

[8] K. R. Chien, M. Shimizu, M. Hoshijima, S. Minamisawa, A. A. Grace Toward molecular strategies for heart disease- Past, Present, Future- *Jpn Circ. J.* 1997; 61:91-118.

[9] S. D. Solomon, A.A.T. Geisterfer-Lowrance, H.P. Vosberg, G. Hiller, J. A. Jarcho, C. C. Mortonj, W.0. McBride, A.L. Mitchell, A.E. Bale, W.J. McKenna, J.G. Seidman, C. E. Seidman A Locus for Familial Hypertrophic Cardiomyopathy Is Closely Linked to the Cardiac Myosin Heavy Chain Genes, CRI-L436, and CRI-L329 on Chromosome 14 at qll-q12 *Am. J. Hum. Genet.* 1990; 47:389-394.

[10] Boerwinkle E, Hixson JE, Hanis CL. Peeking under the peaks: following up genome-wide linkage analyses. *Circulation.* 2000;102:1877–1878.

[11] Yamada Y, Izawa H, Ichihara S, Takatsu F, Ishihara H, Hirayama H, Sone T, Tanaka M, Yokota M. Prediction of the risk of myocardial infarction from polymorphisms in candidate genes. *N. Engl. J. Med.* 2002;347:1916 –1923.

[12] M. Satoda, F. Zhao, G. A. Diaz, J. Burn, J. Goodship, H. R.Davidson, M.E.M.Pierpont, B.D. Gelb Mutations in TFAP2B cause Char syndrome, a familial form of patent ductus arteriosus *Nature Genetics* 2000;25:42-46.

[13] A. Mani, S.M. Meraji, R. Houshyar, J. Radhakrishnan, A. Mani, M. Ahangar, T.M. Rezaie, M.A. Taghavinejad, B. Broumand, H. Zhao, C. Nelson-Williams, and R.P. Lifton *Finding genetic contributions to sporadic disease: recessive locus at 12q24 commonly contributes to patent ductus arteriosus PNAS* 2002; 99 (23):15054–15059.

[14] F. Zhao, C.G. Weismann, M. Satoda, M.E.M. Pierpont, E. Sweeney, E.M. Thompson, B.D. Gelb Novel TFAP2B Mutations That Cause Char Syndrome Provide a Genotype-Phenotype Correlation *Am. J. Hum. Genet.* 2001; 69:695–703.

[15] F. Vitelli, I. Taddei, M. Morishima, E.N. Meyers, E.A. Lindsay, A. Baldini *A genetic link between Tbx1 and fibroblast growth factor signalling Development* 2002; 129: 4605-4611.

[16] H. Yamagishi, J. Maeda, T. Hu, J. McAnally, S.J. Conway, T. Kume, E.N. Meyers, C. Yamagishi, D. Srivastava Tbx1 is regulated by tissue-specific forkhead proteins through a common Sonic hedgehog-responsive enhancer *Genes and Dev.* 2003;17: 269-281.

[17] D.W. Benson, G.M. Silberbach, A. Kavanaugh-McHugh, C. Cottrill, Y. Zhang, S. Riggs, O. Smalls, M.C. Johnson, M.S. Watson, J.G. Seidman, C.E. Seidman, J. Plowden, J.D. Kugler Mutations in the cardiac transcription factor NKX2.5 affect diverse cardiac developmental pathways *J. Clin. Invest.* 1999;104:1567–1573.

[18] F.P. Bernier, R. Spaetgens The geneticist's role in adult congenital heart disease *Cardiol Clin* 2006;24:557-569.

[19] E. Goldmuntz, E. Geiger, D.W. Benson NKX2.5 *Mutations in Patients With Tetralogy of Fallot Circulation* 2001;104;2565-2568.

[20] K. Heathcote, C. Braybrook, L. Abushaban, M. Guy, M. E. Khetyar, M.A. Patton, N. D. Carter, P.J. Scambler. P. Syrris Common arterial trunk associated with a homeodomain mutation of NKX5.6 *Human Molecular Genetics* 2005; Vol. 14, No. 5: 585-93.

[21] A. Pizzuti, A. Sarkozy, A.L. Newton, E. Conti, E. Flex, M.C. Digilio, F. Amati, D. Gianni, C. Tandoi, B. Marino, M. Crossley, B. Dalla Piccola Mutations of ZFPM2/FOG2 Gene in Sporadic Cases of Tetralogy of Fallot *Human Mutation* 2003, 22:372-377.

[22] M.E. Pierpont, C.T. Basson, D.W. Benson, Jr, B.D. Gelb, T.M. Giglia, E. Goldmuntz, G. McGee, C.A. Sable, D. Srivastava, C.L. Webb *Congenital Heart Defects: Current Knowledge: A Scientific Statement From the American Heart Association Congenital Cardiac Defects Committee, Council on Cardiovascular Disease in the Young:* Endorsed by the American Academy of Pediatrics Circulation 2007;115;3015-3038.

[23] D I Wilson, I E Cross, J A Goodship, S Coulthard, A H Carey, P J Scambler, H H Bain, A S Hunter, P E Carter, J Burn DiGeorge syndrome with isolated aortic coarctation and isolated ventricular septal defect in three sibs with a 22q11 deletion of maternal origin *Br. Heart J.* 1991;66:308-12.

[24] F.Greenberg, F.F.B. Elder, P. Haffner, H. Northrup, D. Ledbetter Cytogenetic findings in a prospective series of patients with DiGeorge anomaly *Am. J. Hum. Genet.* 1988;43:605-611 *http://www.geneclinics.org/.*

[25] E.Iwarsson, L.Ahrlund-Richter, J.Inzunza, M.Fridstrom, B.Roselund, T.Hillensjo, P.Sjoblom, M.Nordenskjold, E.Blennow Preimplantation genetic diagnosis of DiGeorge syndrome *Molecular Human Reproduction*, 1998, vol4, no.9 pp. 871-875.

[26] Greenberg F, Elder FF, Haffner P, Northrup H, Ledbetter DH. Cytogenetic findings in a prospective series of patients with DiGeorge anomaly. *Am. J. Hum. Genet.* 1988;43:605– 611.

[27] W.J. Fibison, M. Budarf, H. McDermid, F. Greenberg, and B.S. Emanuel Molecular studies of DiGeorge Syndrome *Am. J. Hum. Genet.* 1990;46:888-895.

[28] H.F. Sutherland, R. Wadey, J.M. McKie, C. Taylor, U. Atif, K.A. Johnstone, S. Halford, U.J. Kim, J. Goodship, A. Baldini, P.J. Scambler Identification of a Novel Transcript Disrupted by a Balanced Translocation Associated with DiGeorge Syndrome *Am. J. Hum. Genet.* 1996, 59:23-31.

[29] W. Gong, B.S. Emanuel, J. Collins, D.H. Kim, Z. Wang, F. Chen, G. Zhang, B. Roe, M.L. Budarf A transcription map of the DiGeorge and velo-cardio-facial syndrome minimal critical region on 22q11 *Human Molecular Genetics* 1996; Vol. 5, No. 6: 789–800.

[30] Deborah A. Driscoll, Marcia L. Budarft and Beverly S. Emanuelt A Genetic Etiology for DiGeorge Syndrome: Consistent Deletions and Microdeletions of 22q11 *Am. J. Hum. Genet.* 1992,50:924-933.

[31] Digilio MC, Angioni A, De Santis M, Lombardo A, Giannotti A, Dallapiccola B, Marino B. Spectrum of clinical variability in familial deletion 22q11.2: from full manifestation to extremely mild clinical anomalies. *Clin. Genet.* 2003;63:308 –313.

[32] Goldmuntz E, Clark BJ, Mitchell LE, Jawad AF, Cuneo BF, Reed L, McDonald-McGinn D, Chien P, Feuer J, Zackai EH, Emanuel BS, Driscoll DA. Frequency of 22q11 deletions in patients with conotruncal defects. *J. Am. Coll Cardiol.* 1998;32:492–498.

[33] Alan J. Mears, Alessandra M. V. Duncan,t Marcia L. Budarf, Beverlr S. Emanuel, Beatrice Sellinger, Jacqueline Siegel-Bartelt, Cheryl R. Greenberg,and Heather E. McDermid Molecular Characterization of the Marker Chromosome Associated with Cat Eye Syndrome *Am. J. Hum. Genet.* 1994, 55:134-142.

[34] Ewart AK, Morris CA, Atkinson D, Jin W, Sternes K, Spallone P, Stock, AD, Leppert M, Keating MT. Hemizygosity at the elastin locus in a developmental disorder, Williams syndrome. *Nat Genet.* 1993;5:11–16.

[35] Hans-Christoph Duba, Andreas Doll, Michael Neyer, Martin Erdel, Christian Mann, Ignaz Hammerer, Gerd Utermann and Karl-Heinz Grzeschik The elastin gene is disrupted in a family with a balanced translocation t(7;16)(q11.23;q13) associated with a variable expression of the Williams-Beuren syndrome *European Journal of Human Genetics* (2002) 10, 351-361.

[36] Luis A. Pérez Jurado, Yu-Ker Wang, Risa Peoples, Antonio Coloma, Jesús Cruces and Uta Francke A duplicated gene in the breakpoint regions of the 7q11.23 Williams–Beuren syndrome deletion encodes the initiator binding protein TFII-I and BAP-135, a phosphorylation target of BTK *Human Molecular Genetics*, 1998, Vol. 7, No. 3, 325-334.

[37] Lucy R. Osborne, Martin Li, Barbara Pober, David Chitayat, Joann Bodurtha, Ariane Mandel, Teresa Costa, Theresa Grebe, Sarah Cox, Lap-Chee Tsui Stephen W. Scherer A 1.5 million–base pair inversion polymorphism in families with Williams-Beuren syndrome *Nature Genetics* volume 29, 2001, 321-325.

[38] *A multifunctional DNA-binding protein that promotes the formation of serum response factor/homeodomain complexes: identity to TFII-I* Dorre A. Grueneberg, R. William Henry, Andrew Brauer, Carl D. Novina, Venugopalan Cheriyath, Ananda L.Roy and Michael Gilman 1997 11: 2482-2493 Genes and Dev.

[39] Bedell MA, Jenkins NA, Copeland NG: Good genes in bad neighbourhoods. *Nat Genet* 1996; 12: 229 ± 232.

[40] Dirk-Jan Kleinjan and Veronica van Heyningen Position effect in human genetic disease *Human Molecular Genetics*, 1998, Vol. 7, No. 10.

[41] Vett K. Lloyd, David Dyment, Donald A.R. Sinclair, and Thomas A. Grigliatti Different patterns of gene silencing in position-effect variegation *Genome* 46: 1104–1117, 2003.

[42] Mayada Tassabehji, Martin Carette, Carrie Wilmot, Dian Donnai, Andrew P Read and Kay Metcalfe A transcription factor involved in skeletal muscle gene expression is deleted in patients with Williams syndrome *European Journal of Human Genetics* (1999) 7, 737–747.

[43] Nancy B. Spinner, Elizabeth B. Rand, Paolo Fortina, Anna Genin, Rebecca Taub, Antonio Semeraro, and David A. Piccoli Cytologically Balanced t(2;20) in a Two-Generation Family with Alagille Syndrome: Cytogenetic and Molecular Studies *Am. J. Hum. Genet.* 55:238-243, 1994.

[44] K. M. Loomes, L. A. Underkoffler, J. Morabito, S. Gottlieb, D. A. Piccoli, N. B. Spinner, H.S. Baldwin, and R. J. Oakey The expression of Jagged1 in the developing mammalian heart correlates with cardiovascular disease in Alagille syndrome *Human Molecular Genetics*, 1999, Vol. 8, No. 13:2443-2449.

[45] Elizabeth B. Rand, Nancy B. Spinner, David A. Piccoli, Peter F. Whitington, and Rebecca TaubMolecular Analysis of 24 Alagille Syndrome Families Identifies aSingle Submicroscopic Deletion and Further Localizes the Alagille Region within 20p12 *Am. J. Hum. Genet.* 57:1068-1073, 1995.

[46] Albrecht Röpke, Annegret Kujat, Mechthild Gräber, Joannis Giannakudis, and Ingo Hansmann Identification of 36 Novel Jagged1 (JAG1) Mutations in Patients With Alagille Syndrome *Human Mutation* 2003 Jan;21(1):100.

[47] E A Jones,M Clement-Jones, D I Wilson JAGGED1 expression in human embryos: correlation with the Alagille syndrome phenotype *J. Med. Genet.* 2000;37:658–662.

[48] Raymond P. Colliton, Lynn Bason, Feng-Min Lu, David A. Piccoli, Ian D. Krantz and Nancy B. Spinner Mutation Analysis of Jagged1 (JAG1) in Alagille Syndrome Patients *Human Mutation* 2001 Feb;17(2):151-2.

[49] Ian D. Krantz,1 Raymond P. Colliton, Anna Genin, Elizabeth B. Rand, Linheng Li, David A. Piccoli, and Nancy B. Spinner Spectrum and Frequency of Jagged1 (JAG1) Mutations in Alagille Syndrome Patients and Their Families *Am. J. Hum. Genet.* 62:1361–1369, 1998.

[50] J. J. D. Morrissette, R. P. Colliton and N. B. Spinner Defective intracellular tran sport and processing og JAG1 missense mutations in Alagille syndrome *Human Molecular Genetics,* 2001, Vol. 10, No.14.

[51] B M Kamath, L Bason, D A Piccoli, I D Krantz, N B Spinner Consequences of JAG1 mutations *J. Med. Genet.* 2003;40:891–895.

[52] Binita M. Kamath, Ian D. Krantz, Nancy B. Spinner, James E. Heubi, and David A. Piccoli Monozygotic Twins With a Severe Form of Alagille Syndrome and Phenotypic Discordance*American Journal of Medical Genetics* 112:194–197 (2002)

[53] Ayala Laufer-Cahana, Ian D. Krantz, Lynn D. Bason, Feng-Min Lu, David A. Piccoli, and Nancy B. Spinner Alagille Syndrome Inherited From a Phenotypically Normal Mother With a Mosaic 20p Microdeletion *American Journal of Medical Genetics* 112:190–193 (2002).

[54] Ryan McDaniell, Daniel M. Warthen, Pedro A. Sanchez-Lara, Athma Pai, Ian D. Krantz,David A. Piccoli, and Nancy B. Spinner NOTCH2 Mutations Cause Alagille Syndrome, a Heterogeneous Disorder of the Notch Signaling PathwayThe American *Journal of Human Genetics Volume 79* July 2006.

[55] Marco Tartaglia, Kamini Kalidas, Adam Shaw, Xiaoling Song, Dan L. Musat, Ineke van der Burgt, Han G. Brunner, Debora R. Bertola, Andrew Crosby, Andra Ion, Raju S. Kucherlapati, Steve Jeffery, Michael A. Patton, and Bruce D. Gelb PTPN11 Mutations in Noonan Syndrome: Molecular Spectrum, Genotype-Phenotype Correlation, and Phenotypic Heterogeneity *Am. J. Hum. Genet.* 70:1555–1563, 2002.

[56] Ineke van der Burgt Noonan syndrome Orphanet *Journal of Rare Diseases* 2007, 2:4

[57] Tartaglia M, Kalidas K, Shaw A, Song X, Musat DL, van der Burgt I,Brunner HG, Bertola DR, Crosby A, Ion A, Kucherlapati RS, Jeffery S, Patton MA, Gelb BD. PTPN11 mutations in Noonan syndrome: molecular spectrum, genotype-phenotype correlation, and phenotypic heterogeneity. *Am. J. Hum. Genet.* 2002;70:1555–1563.

[58] Rie Yoshida, Tomonobu Hasegawa, Yukihiro Hasegawa, Toshiro Nagai, Eiichi Kinoshita, Yoko Tanaka, Hirokazu Kanegane, Kenji Ohyama, Toshikazu Onishi, Kunihiko Hanew, Torayuki Okuyama, Reiko Horikawa, Toshiaki Tanaka, and Tsutomu Ogata Protein-Tyrosine Phosphatase, Nonreceptor Type 11Mutation Analysis and Clinical Assessment in 45 Patients with Noonan Syndrome *The Journal of Clinical Endocrinology and Metabolism* 89(7):3359–3364.

[59] Digilio MC, Conti E, Sarkozy A, Mingarelli R, Dottorini T, Marino B, Pizzuti A, Dallapiccola B. Grouping of multiple-lentigines/LEOPARD and Noonan syndromes on the PTPN11 gene. *Am. J. Hum. Genet.* 2002; 71:389 –394.

[60] Anna Sarkozy Maria Cristina Digilio Bruno Dallapiccola LEOPARD syndrome Orphanet *Journal of Rare Diseases* 2008, 3:13.

[61] Giuseppe Limongelli, Anna Sarkozy, Giuseppe Pacileo, Paolo Calabro`, Maria Cristina Digilio, Valeria Maddaloni, Giulia Gagliardi, Giovanni Di Salvo, Maria Iacomino, Bruno Marino, Bruno Dallapiccole, Raffaele Calabro' Genotype–Phenotype Analysis and Natural History of Left Ventricular Hypertrophy in LEOPARD Syndrome *American Journal of Medical Genetics* 2008 Mar 1;146A(5):620-8.

[62] Limongelli G, Pacileo G, Marino B, Digilio MC, Sarkozy A, Elliott P, Versacci P, Calabro P, De Zorzi A, Di Salvo G, Syrris P, Patton M, McKenna WJ, Dallapiccola B, Calabro R. Prevalence and clinical significance of cardiovascular abnormalities in patients with the LEOPARD syndrome *Am. J. Cardiol.* 2007 Aug 15;100(4):736-41. Epub 2007 Jun 27.

[63] Maria Cristina Digilio , Emanuela Conti, Anna Sarkozy, Rita Mingarelli, Tania Dottorini, Bruno Marino, Antonio Pizzuti, and Bruno Dallapiccola Grouping of Multiple-Lentigines/LEOPARD and Noonan Syndromes on the PTPN11 *Gene Am. J. Hum. Genet.* 71:389–394, 2002.

[64] Digilio MC, Sarkozy A, de Zorzi A, Pacileo G, Limongelli G, Mingarelli R, Calabrò R, Marino B, Dallapiccola B. LEOPARD syndrome: clinical diagnosis in the first year of life. *Am. J. Med. Genet. A.* 2006 Apr 1;140(7):740-6.

[65] Deborah A. Mcdermott, Michael C. Bressan, Jie He, Joseph S. Lee, Salim Aftimos, Martina Brueckner, Fred Gilbert, Gail E. Graham, Mark C. Hannibal, Jeffrey W. Innis, Mary Ella Pierpont, Annick Raas-Rothschild, Alan L. Shanske, Wendy E. Smith, Robert H. Spencer, Martin G. St. John-Sutton, Lionel van Maldergem, Darrel J.

Waggoner, Matthew Weber, and Craig T. Basson TBX5 *Genetic Testing Validates Strict Clinical Criteria for Holt-Oram Sindrome Pediatric Research Vol. 58*, No. 5, 2005.

[66] Hiroi Y, Kudoh S, Monzen K, Ikeda Y, Yazaki Y, Nagai R, Komuro I. Tbx5 associates with Nkx2-5 and synergistically promotes cardiomyocyte differentiation. *Nat. Genet.* 2001;28:276 –280.

[67] Dixon JW, Costa T, Teshima IE. Mosaicism for duplication 12q (12q13–q24.2) in a dysmorphic male infant. *J. Med. Genet.* 1993;30: 70–72.

[68] McCorquodale MM, Rolf J, Ruppert ES, Kurczynski TW, Kolacki P. Duplication (12q) syndrome in female cousins, resulting from maternal (11;12) (p15.5;q24.2) translocations. *Am. J. Med. Genet.* 1986;24:613–622.

[69] Robert G. Resta Defining and Redefining the Scope and Goals of Genetic Counseling *American Journal of Medical Genetics Part C* (Seminars in Medical Genetics) 142C:269–275 (2006).

[70] Angus Clarke, Martin Richards, Lauren Kerzin-Storrar, Jane Halliday, Mary Anne Young, Sheila A Simpson, Katie Featherstone, Karen Forrest, Anneke Lucassen, Patrick J Morrison, Oliver WJ Quarrell, Helen Stewart and collaborators Genetic professionals' reports of nondisclosure of genetic risk information within families *European Journal of Human Genetics* (2005) 13, 556–562.

[71] Pascal Borry, Jean-Pierre Fryns, Paul Schotsmans and Kris Dierickx Carrier testing in minors: a systematic review of guidelines and position papers *European Journal of Human Genetics* (2006) 14, 133.

Index

B

C

D

E

F

G

H

N

O

Q

R

S